Threatcasting

Synthesis Lectures on Threatcasting

Editors
Brian David Johnson, *Arizona State University*
Natalie Vanatta, *United States Military Academy*

Threatcasting
Brian David Johnson, Natalie Vanatta, and Cyndi Coon
2021

Threatcasting
Brian David Johnson, Natalie Vanatta, and Cyndi Coon

ISBN: 978-3-031-01447-5 Paperback
ISBN: 978-3-031-02575-4 PDF
ISBN: 978-3-031-00345-5 Hardcover

DOI 10.1007/978-3-031-02575-4
A Publication in the Springer series

SYNTHESIS LECTURES ON THREATCASTING
Lecture #01
Series Editors: Brian David Johnson, Arizona State University and Natalie Vanatta, United States Military Academy
Series ISSN: [Pending] Print   [Pending] Electronic

Threatcasting

Brian David Johnson
Arizona State University

Natalie Vanatta
United States Military Academy

Cyndi Coon
Laboratory5

SYNTHESIS LECTURES ON THREATCASTING #01

ABSTRACT

Threatcasting uses input from social science, technical research, cultural history, economics, trends, expert interviews, and even a little science fiction to recognize future threats and design potential futures. During this human-centric process, participants brainstorm what actions can be taken to identify, track, disrupt, mitigate, and recover from the possible threats. Specifically, groups explore how to transform the future they desire into reality while avoiding an undesired future. The Threatcasting method also exposes what events could happen that indicate the progression toward an increasingly possible threat landscape.

This book begins with an overview of the Threatcasting method with examples and case studies to enhance the academic foundation. Along with end-of-chapter exercises to enhance the reader's understanding of the concepts, there is also a full project where the reader can conduct a mock Threatcasting on the topic of "the next biological public health crisis." The second half of the book is designed as a practitioner's handbook. It has three separate chapters (based on the general size of the Threatcasting group) that walk the reader through how to apply the knowledge from Part I to conduct an actual Threatcasting activity. This book will be useful for a wide audience (from student to practitioner) and will hopefully promote new dialogues across communities and novel developments in the area.

Impending technological advances will widen an adversary's attack plane over the next decade. Visualizing what the future will hold, and what new threat vectors could emerge, is a task that traditional planning mechanisms struggle to accomplish given the wide range of potential issues. Understanding and preparing for the future operating environment is the basis of an analytical method known as Threatcasting. It is a method that gives researchers a structured way to envision and plan for risks ten years in the future.

KEYWORDS

Threatcasting, strategic foresight, future, strategy, strategic planning, risk assessment, futurism, long term strategic foresight, dystopian futures, applied futurism, framework foresight, science fiction prototyping, participatory design

Contents

Foreword by Andy Hines

Thank goodness we have futurists such as Brian David Johnson (BDJ) who are willing to stop, write down, and share with us their tremendous work. Like BDJ, I came to academia after a lengthy career in the professional space focusing on applied futures. I can attest to the challenge of making the time to catalog the work, given the constant pressure to move on to the next project. One of my major laments about the field is that it is practitioner-heavy, and practitioners are often too busy to catalog their work. I recall hearing about BDJ's early work at Intel. I was grateful that he published a key piece of that work in his book *Science Fiction Prototyping: Designing the Future with Science Fiction*. Then I started hearing about his new work, *Threatcasting*, which I have the pleasure of introducing here.

I recall the University of Hawaii's legendary futurist Jim Dator's concern about a decade ago that there hadn't been much new happening on the methodology front—except for CLA. I felt like there was some interesting, applied futures work happening, but that in some cases it was more tweaks than inventions, or for proprietary reasons that work simply was not getting reported. I eagerly jumped at the opportunity to see for myself in 2018, when I took on the methodology section of the *Knowledge Base 2020* update with Richard Slaughter. This gave me the opportunity to look for what had been happening over the last 15 years. As I suspected, a little digging revealed methodological innovation was indeed alive and well. Not surprisingly, I came across BDJ's work on Science Fiction Prototyping which we quickly snapped up and included in that volume. In conversations with BDJ related to that work, I learned more about Threatcasting.

It's one thing to be excited by the simple act of sharing; it's another to be excited by what is being shared. And I am excited by what is being shared. I have half-jokingly referred to myself as a practitioner in academics' clothing now that academic is my full-time gig. BDJ is moving in this very same direction, and I suspect may fit a similar description. This is great news for those of us who want a practical hands-on guide to doing the work.

As I reviewed the *Threatcasting* method, the approach immediately resonated with the Framework Foresight approach we teach at the University of Houston. I promise my students that they will learn to "diagnose" how any applied method explores the future, provided the method follows sound practice. A key aspect of that is how the methods align with the key foresight competencies that were identified in the APF Competency model work published back in 2017. In other words, a sound method is grounded and aligned with key foresight competencies. That is the case here with *Threatcasting*. At the same time, the method has BDJ's "special sauce." This to me is precisely what the field needs more and more of—aligned and grounded applied foresight methods that bring some new ideas and innovation. This, in my opinion, is how you build a field: new contributions that build upon and enhance a growing body of work.

The book details the nuts and bolts of the *Threatcasting* method. Those nuts and bolts are also contextualized and illustrated in a way that drive homes not just the "what to do" but also the "why do it." Again, this resonates with what we preach here at the University of Houston Foresight program. We want to prepare people to do foresight work in a practical sense, but also know why they are doing what they are doing. I think BDJ would wholeheartedly agree that we are preparing chefs, not cooks.

Any doubts about what you are getting with *Threatcasting* should be quickly erased by the logo of the Threatcasting Lab at Arizona State University: "Envisioning Futures to Empower Action." It addresses the *what* we are doing—envisioning futures—and *why* are we are doing it: to empower action. Exactly!

The tone of the book is quite appealing. This is not one of those books written by a smart person who wants to convince you how smart they are. BDJ and his co-authors are very smart people who come off as people you'd like to have a beer or coffee with—people who are trying to help you learn. The entire book is structured and reads as if it was written by authors coming from the customer/reader perspective. How will they see this? How will they use it? Will this be clear to them? Those seem to me to be the types of questions that are being asked and are addressed in this book.

Let me share a few features I really like.

- **Actual project examples:** One of the cool things I like about what we are doing at Houston Foresight is that we get students involved in "real" paid project work. What better way to learn than to apply what you're working with in the classroom on a real project? I see this same idea in play here in *Threatcasting*. A key aspect of teaching the methodology is by working it in real projects. Not only is this practical experience priceless, but it also provides a platform to continually test and improve the method. I suspect we will see a second edition at some point.

- **Exercises:** The are exercises sprinkled throughout the work that give the reader a chance to try it out. I suspect many readers will be using it as a text, so having exercises is really helpful. Along those lines, the entire text is written in a way that will help teachers teach.

It's exciting to see leading practitioners openly sharing their work. It speaks to collegiality in the field and to the growing demand for foresight. Practitioners will find plenty to add to their tool kit, clients will find an exciting new way to explore the future, and we professors will have more good material teach. I could go on, but now I think it's time to pass the baton to BDJ, Natalie and Cyndi.

– Andy Hines, Associate Professor, University of Houston Foresight Program, July 4, 2021

Preface

Threatcasting is a method that enables multidisciplinary groups to envision and plan systematically against threats ten years in the future. Groups explore how to transform the future they desire into reality while avoiding an undesired future. Threatcasting uses inputs from social science, technical research, cultural history, economics, trends, expert interviews, and even a little science fiction. These various inputs allow the creation of potential futures (focused on the fiction of a person in a place doing a thing). Some of these futures are desirable while others are to be avoided. By placing the threats into a fictional story, it allows decision makers and practitioners to imagine what needs to be done today, as well as four and eight years into the future, to empower or disrupt the targeted future scenario. The framework also illustrates what flags, or warning events, could appear in society that indicate the progress toward the threat future.

In the early 2000's, Brian David Johnson (BDJ) was the chief futurist for the Intel Corporation. As a high-tech manufacturing firm, it took them about ten years to design, develop, and deploy a microprocessor. Therefore, it was of vital business importance for them to know what people would want to do with computers ten years in the future. As an applied futurist, he developed the necessary steps to move toward positive futures and away from the negative.

Early on, the company also realized that you could use the output from Brian's work for more than just silicon chip design It was used to explore, plan, and train the workforce the company would need to build the chip, develop the software to run on it, and develop the ecosystem of companies to make use of all the emerging capabilities of these new computational platforms. It illuminated what kind of partners might be needed to make the product successful. It showed what new capabilities would be needed (some of which were developed internally and, for others, this work served as guidance for possible mergers or acquisitions by the company). In fact, a number of patents were written based on the work. Ultimately, the marketing and PR team really liked the work because having a vision for the future can not only guide the company it can also capture the imagination of the consumer and get them excited for how it could impact their lives. The Threatcasting Method was created out of necessity. It solved a basic problem that can be applied to multiple industries and application areas: How do we envision the futures we don't want and then systematically work to disrupt, mitigate, or recover from them? The Method identified a range of negative futures and then allowed the organization to track the progress of these emerging threats and begin to take specific, measurable steps to stop them from occurring.

Proving its efficacy at the Intel Corporation and in Silicon Valley, other organizations began to adopt and adapt the method to solve their own problems. The Method was refined due to collaboration with organizations like the United States Air Force Academy, where we modeled early explorations of cyber warfare and lethal autonomous devices, as well as nuclear proliferation. Also, with the Federal Emergency Management Agency (FEMA) we explored the future of wildfires and earthquakes, understanding that while we couldn't prevent these from happening, we could plan for their mitigation and recovery. In 2017, the Threatcasting Lab at Arizona State University was established to continue to develop the Method, convene workshops, and train practitioners and students.

This book is broken into two parts. Part 1 will give you an overview of the Threatcasting concept with examples and case studies to enhance the academic foundation. At the end of these chapters, you will find exercises to enhance your understanding of the Method and prepare you to apply what you learned. Part 1 ends with a project chapter where you can conduct a mock Threatcasting on the topic of "the next biological public health crisis." This will enable you to pull together the knowledge from Part 1 and apply it.

Part 2 is designed as a practitioner's handbook Chapters 9, 10, and 11 describe the three general sizes of groups with which you might want to perform a Threacasting activity. You can either read these chapters synchronously to understand similarities and differences between these situations or read them independently as a direct reference (i.e., handbook) when you are going to run a specific threatcasting session.

To illustrate the key points within the book, the following various ingredients are used:

- **Interviews** with participants, subject matter experts, facilitators, curators, writers, academics, artists, advertising agents, and other government and military officials are provided to offer a lens into the many different critical voices that are necessary to create Threatcasting results. They each have unique experiences with the process, and their stories are shared to enrich the text.

- **Stories from the Lab** are snippets of stories derived from conducting various Threatcasting activities over the years. The purpose of these stories is to highlight lessons learned and funny situations that have occurred during the application of the Threatcasting Method to a multitude of topics for various organizations over the years.

- **Exercises** give the opportunity to pause and work on building your threatcasting muscles before proceeding to the next chapter. Some of the exercises are considered "skill building"—focused on developing a skill or concept that will be useful when Threatcasting. Other exercises provide an opportunity to experience an aspect of the Threatcasting Method in an applied space.

- **Project** is a mock Threatcasting from beginning to end for the purposes of practicing and learning the Method. The project will contain all the elements in Part 1 so that the analyst(s) can rehearse, experiment, and practice the Threatcasting Method in a contained environment.

How To Get the Most From This Book

This book is broken into two parts. Part 1 will give you an overview of the Threatcasting concept with examples and case studies to enhance the academic foundation. At the end of these chapters, you will find exercises to enhance your understanding of the method and prepare you to apply what you learned. Part 1 ends with a project chapter where you can conduct a mock Threatcasting on the topic of "the next biological public health crisis." This will enable you to pull together the knowledge from Part 1 and apply it.

Part 2 of the book is designed as a practitioner's handbook. Chapters 9, 10, and 11 describe the three general sizes of groups that you might want to perform a Threacasting activity with. You can either read these chapters synchronously to understand similarities and differences between these situations, or you can read them independently as a direct reference (i.e., handbook) when you are going to run a specific threatcasting session.

Interview

Throughout this book, you will find Interviews. These are conversations with participants, subject matter experts, facilitators, curators, writers, academics, artists, advertising agents, and other government and military officials. These interviews are provided to offer a lens into the many different critical voices that are necessary to create Threatcasting results. They each have unique experiences with the process, and their stories are shared in order to enrich the text.

Conversations in the Lab

Similar to the interviews, conversations between members of the lab are captured to provide an applied understanding of the intended use of elements in the Threatcasting foundation. These conversations are shared in order to enrich the text.

Exercise

At the end of the Part 1 chapters, you will find Exercises. The purpose of providing exercises is to pause and work on building your threatcasting muscles before proceeding to the next chapter. Some of the exercises are considered "skill building"—focused on developing a skill or concept that will be useful when Threatcasting. Other exercises provide an opportunity to experience an aspect of the Threatcasting Method in an applied space. Each time you find an exercise, plan to run through it yourself or with a group, capture notes, and reference them when you are building your own threatcasting models. Completing the exercises in order will help build the skills needed for the next chapter and the next.

Project

At the end of Part 1, there is a project. This project is a mock Threatcasting from beginning to end for the purposes of practicing and learning the Method. The project will contain all the elements in Part 1 so that the analyst(s) can rehearse, experiment, and practice the Threatcasting Method in a contained environment.

Stories from the Lab

As the interviews enhanced the foundational concepts in Part 1, throughout Part 2 there are snippets of stories from the Threatcasting Lab. These stories are derived from conducting various Threatcasting activities over the years. The purpose of these stories is to highlight lessons learned and funny situations that have occurred during the application of the Threatcasting Method to a multitude of topics for various organizations over the years.

LET'S GET STARTED!

The best way to get to know and use Threatcasting is to conduct Threatcasting sessions. It's like fishing: the first time you go out, someone has to show you how to put the three pieces of the pole together, attach the reel, thread the line, attach the hook and worm, and how to cast. Then there is a whole conversation about what to do while the worm is in the water. You need to fish a few times to get the hang of it. Generally, each time the pieces are the same and go together the same, but the outcomes could leave you with a big fish story!

In the second part of the book, we will show you how to apply Threatcasting to multiple situations, including running a session or workshop by yourself, in a small group, or in a large group. Each of these types of sessions have pros and cons. The goal of the book is to get you to start Threatcasting right away.

Threatcasting is a method with two high-level segments. The first segment imagines possible futures by gathering together a broad range of multidisciplinary inputs to model a range of possible and potential threats 10 years in the future. The second segment then backcasts—specifying what needs to happen over the next 10 years to monitor, disrupt, mitigate, and/or recover from these threats. As Threatcasting is an applied futures methodology, these actions are specific to an organization, group of people, or research agenda.

This book combines method explanation (Part 1) and a "how to" section (Part 2) so that you can apply it to specific problems.

Key Terms

This book will use specific terms and descriptors to outline Threatcasting.

The Use of Parenthetical Plurals (s)

When specific people are referenced in the book, most will be followed by an (s) symbol or parenthetical plural. This is done to show that the task can be performed by a single person or multiple people.

Analyst(s)

The analyst(s) role in Threatcasting is to lead the effort. Typically, the lead analyst will run the effort from definition, planning, curation, execution, post-analysis, and reporting. Other analysts can be brought in to support the lead analyst in some or all tasks. These support analyst(s) can be a benefit if the Threatcasting is larger in scope and size. Specifically in the post-analysis stage, support analysts can be brought in to augment and broaden the lead analyst's perspective.

Participant(s)

The participant(s) are brought in by the analyst(s) to take part in the Threatcasting Workshop. Participant(s) are typically involved in the project only during the workshop. However, some participant(s) might also take part in a peer review of the final findings after the analyst(s) conduct the post analysis.

Subject Matter Expert (SME)

A Subject Matter Expert (SME) is a person with a particular expertise, perspective, or opinion that the analyst(s) selects as a prompt for the Threatcasting Workshop. Generally, SMEs do not participate in the workshop. They can, however, participate in a peer review of the final findings after the analyst(s) conduct the post analysis.

Threatcasting Foundation

The Threatcasting Foundation consists of the primary topic, research question, and applications area that the Threatcasting Method will cover. It also lays out how the output will be used. This is discussed in Chapter 2. Establishing a robust, researched, and focused foundation is an important skill for any research project. The better defined the foundation, the more efficient and effective the application of the Threatcasting Method will be.

Science Fiction Prototype

Science Fiction Prototypes (SFP) are plausible science-fiction futures based on the Threatcasting research that allow the participant(s) and reader to explore the ethical, cultural, policy, and security impacts on people. The specific events depicted exist only in the imagination of the authors. The locations are selected merely to dramatize the storylines and could just as easily be any city or nation confronting external threats.

These SFPs are used as a part of the Effect-Based Model (EBM), to give participant(s) a process that elicits details about the future

Effect-Based Model (EBM)

The central device used in the Threatcasting Method is the Effect-Based Model (EBM). The EBM is a qualitative model wherein participant(s) use the prompts and the Research Synthesis Workbook (RSW) data to imagine possible and potential threats in the future.

An EBM is a structured database (e.g., spreadsheet) that poses to the practitioners a series of questions and requires a list of tasks to fill in to create the qualitative threat model. Moving from the high-level research (prompts) and participant(s)' perspectives in the RSW, practitioner(s) focus on a specific threat future. Moving from the macro-level to the micro-level, practitioner(s) explore a person in a place experiencing a threat.

The EBM uses Experience Design Methods and Effects-Based Operations methods, tools, and approaches to guide the participants through the futurecasting and backcasting exercises.

Project Documentation

Project documentation is a collection of files that analyst(s) use throughout the Threatcasting Method. These types of documents can be text docs (e.g., MS Word, Notepad, Google Docs), spreadsheets (e.g., MS excel, Google Sheets), or slides (e.g., MS Powerpoint, Slides). The type of document doesn't matter so much.

The analyst(s) uses the project documentation to keep track of every step on the Threatasting process such as:

- capturing the Threatcasting Foundation;

- generating a list of possible prompts and Subject Matter Experts;

- listing out possible participant(s);

- generating the various workbooks; and

- capturing analyst(s)' notes during post analysis.

Proper and diligent project documentation will allow the process to go smoothly, ease communication with others, simplify tracking, and ensure there is complete data transparency at the end of the process.

Why Poetry Matters

The Arizona State Threatcasting Lab was launched on Valentine's Day 2017.

As a general rule, each Threatcasting project has a poem (sometimes more than one) used to kick off major meetings. We do this for a few reasons. First, it's unexpected. When you arrive in the desert to one of our events, expecting to explore threats a decade into the future, you don't expect to be met with poetry. Unexpectedly hearing poetry makes people a little uncomfortable and takes them out of their normal day-to-day routines. That's an important part of Threatcasting. To get people thinking ten years into the future, we need to leave the present behind.

The T.S. Eliot poem "Burnt Norton" was the first one we ever used.

Burnt Norton

Time present and time past
Are both perhaps present in time future,
And time future contained in time past.
If all time is eternally present
All time is unredeemable.
What might have been is an abstraction
Remaining a perpetual possibility
Only in a world of speculation.
What might have been and what has been
Point to one end, which is always present.
Footfalls echo in the memory
Down the passage which we did not take[1]

T.S. Elliot

Why This Poem Matters

I have been using the T.S. Eliot poem "Burnt Norton" for many years when I kick off Threatcasting sessions. It is my go-to poem, especially when I can't find another that is more appropriate. That's always the goal: kick off a Threatcasting Workshop or "big meeting" with a poem that says something about the lead analyst's motivations for the project or that highlights something about the project in general.

Written in 1935, this passage was part of a larger work called the "Four Quartets." The poem focuses on time: time past, present, and future. It gets the reader to consider the qualities of each.

---

[1] Excerpted from Eliot, T.S. (1936) "Burnt Norton," Four Quartets.

It also explores the different possibilities of the past, present, and future. It races around and plays with time in a way that pushes the reader to think differently about time itself.

Finally, the verse ends by bringing us back to the people. "Footfalls echo in the memory Down the passage we did not take." The metaphysical conversation about time becomes about the humans experiencing that time because they are the center in the experience of it.

Eliot's verse captures many important aspects of Threatcasting. The understanding that time is not fixed, that time can be discussed, and that ultimately all time is about people helps to center us on some of the core values of Threatcasting and this applied futures way of thinking.

The lab's Chief of Staff opens with this verse from Roald Dahl as a reminder that our gift of imagination is what allows us to step into the future.

And above all,
watch with glittering
eyes the whole
world around you
because the greatest
secrets are always
hidden in the most
unlikely places
Those who don't
believe in magic
will never find it.[2]

Roald Dahl

Why This Poem Matters

Dahl's poem reminds the participant(s) and ourselves that Threatcasting is all about people. Everything we do in life is about our fellow humans; it begins and ends with people. There might be some technology, companies, processes, and procedures in between. But ultimately it is a focus on humans and making them safer and the resulting future better for all.

There is magic to be found in the interactions between people, in the conversations and arguments and even in the tense silences. Threatcasting is about people and it's also about people

---

[2] Excerpted from Dahl, R. (1991). *Billy and the Minpins.*

working together and being open to new ideas. Dahl challenges everyone who participates in Threatcasting to believe in the magic of humans so that we may then find it.

This final piece of verse is from the lab's senior advisor. It helps us remember that it is also extremely important when you are Threatcasting to have a sense of humor. Chumbawamba is an English rock band that had a very popular song in the late 1990s.

Tubthumping

I get knocked down, but I get up again
You are never gonna keep me down[3]

Chumbawamba

Why These Lyrics Matter

Lyrics are poems for the soul. I am not a poetry expert—in fact, the last time I probably read and analyzed poetry was in college for a Humanities requirement. Instead, what speaks to me are song lyrics … not those catchy tunes that draw your ear in, but the lyrics that speak to your heart.

Ignoring the fact that a Rolling Stones poll[4] names this song in the top 10 worst of the 1990s, the lyrics remind us all to be resilient and that we have the power to overcome our threat-filled visions of the future. Over the course of a Threatcasting project, you will see the participant(s) achieve highs and lows in their emotional state as they can be overwhelmed in the grim futures that they envision. They especially start to feel horrible for the person they have created who has to live in that world. Having a reminder about the innate resiliency of humankind is important—to know that we are not just going to picture the horrible side of life, but that we are also going to develop a path for that future never to happen.

It is not just participant(s) but also the analyst(s) that need to be reminded of this lesson—that it is possible to get back up. When performing the post-analysis activities, it is easy to get low and feel there is no way past these challenges. I like to play loud and inspiring music during post-analysis to remind myself that nothing is going to keep us down—we will find a way to succeed, and this work is going to help us all.

---

[3] Read the full lyrics at https://bit.ly/3zW5SPI.
[4] https://www.rollingstone.com/music/music-lists/readers-poll-the-worst-songs-of-the-nineties-11176/

Part 1

Threatcasting:

Method, Framework, and Process

CHAPTER 1

Threatcasting

The past, present, and future

Brian David Johnson

It was a cold, clear, winter day. Walking the ground of the United States Military Academy—West Point is a heady experience. Strategically perched at a bend in the Hudson River, the military fort was integral to the then fledgling United States' victory over the British in the Revolutionary war. When you walk the winding paths and enter the solid stone buildings, you know you are walking where George Washington walked. When you look down over the river you realized this is where General Patton did the same thing. Everywhere you go on the campus at West Point you are walking through both the present and the past.

For me this is particularly disorienting. I'm a futurist. I work with organizations to model the future. It's been said I live my life 10 years in the future and commute home on the weekends. The reason I was at West Point was to model the future. In particular, I was working to model potential threats to the United States' national security. In effect, on that winter day I was walking through the past, present, and future.

Walking up the steps of the West Point library, I was returning to the space where a few months before I had led a team of soldiers, security professionals, and even a science fiction comic book writer through a threatcasting session to explore the future of cyber warfare. At the top floor of the library there is a large, open, event space surrounded by windows that overlook the campus. At the top of the steps I ran into General Rhett Hernandez (ret.)

"BDJ," he hollered in his big deep voice. Everyone calls me BDJ. General Hernandez is a towering figure both physically and in his accomplishment. Among many of those accomplishments is that he was the first commander of US Army Cyber Command until he retired. But Generals really don't retire. He remained active and I was fortunate enough to have him as a part of the Threatcasting session.

"Hello, sir," I replied.

"Call me Rhett," he smiled, shaking my hand.

"Yes, sir," I replied.

"Hey, did you see?" he started. His eyes lit up with excitement as he explained some breaking international news related to a new technology.

"I did see that," I replied.

"But that's exactly the thing we talked about in the threatcasting session months ago," he continued. "We said it could happen."

"Yes sir, that's what we do."

He shook his head as if I wasn't understanding. "No, the thing is, we know what to do. It wasn't a surprise. We have a plan."

"Yes sir, that's why we do threatcasting," I smiled.

"That's great," he slapped me on the shoulder. "I guess you're right. Now come with me there's a person I want you to brief…"

1.1   THREATCASTING OVERVIEW

Threatcasting is a method that enables multidisciplinary groups to envision and plan systematically against threats ten years in the future. Groups explore how to transform the future they desire into reality while avoiding an undesired future. Threatcasting uses inputs from social science, technical research, cultural history, economics, trends, expert interviews, and even a little science fiction. These various inputs allow the creation of potential futures (focused on the fiction of a person in a place doing a thing). Some of these futures are desirable while others are to be avoided. By placing the threats into a fictional story, it allows decision makers and practitioners to imagine what needs to be done today as well as four and eight years into the future to empower or disrupt the targeted future scenario. The framework also illustrates what flags, or warning events, could appear in society that indicate the progress toward the threat future.

Threatcasting is a human-centric process, and therefore the humans that participate in a threatcasting session are critical. Regardless of age, experience, or education, all participants are considered practitioners. Threatcasting is a theoretical exercise undertaken by practitioners with special domain knowledge of how to specifically disrupt, mitigate, and recover from theoretical threat futures. Additionally, a few participants are curated to be outliers, trained foresight professionals, and young participants for a fresh and multi-generational perspective in the groups. When using threatcasting on military problems, the mixture of participants should span academia, private industry, government, and the military.

1.2   THREATCASTING OVERVIEW: AN ONTOLOGICAL DISCUSSION

To introduce Threatcasting it might be helpful to have an ontological discussion. When it comes to information science, an "ontology encompasses a representation, formal naming, and definition of the categories, properties, and relations between the concepts, data, and entities that substantiate one, many, or all domains of discourse. More simply, an ontology is a way of showing the properties of a subject area and how they are related, by defining a set of concepts and categories that represent the subject." (Ontology, 2021)

For our Threatcasting discussion, we are going to review what Threatcasting is. For our purposes, Threatcasting is a method, a framework, and a process. By exploring how Threatcasting has the properties of each of these, we can get a better high-level understanding of it before we dig deeper into the details.

1.2.1   THREATCASTING AS A METHOD

"A method is simply a research tool, a component of research - say for example, a qualitative method such as interviews. Methodology is the justification for using a particular research method."—Deborah Gabriel (2011)

"A methodology is an approach to "doing something" with a defined set of rules, methods, tests activities, deliverables, and processes which typically serves to solve a specific problem...Methodologies demonstrate a well thought out, defined, repeatable approach."—Scott Ellis (2008)

Threatcasting as a method provides analyst(s) with a series of defined steps, tasks, systems, and ultimately outputs to identify a range of possible and potential threats, along with actions to be taken and indicators to be monitored. Threatcasting is an applied futures or foresight method that not only identifies a range of possible and potential futures but specifies indicator(s) and action(s) for a specific audience. Meaning that Threatcasting Projects are conducted for a specific group or organization so that they can put them to use. Threatcasting identifies these possible and potential futures for a specific organization so that they can take action.

Threatcasting is an analytical method that explores tomorrow's threats today, in order to give organizations and communities time to detect, prepare, disrupt, mitigate, and, when needed, recover. The goal of Threatcasting is not to predict the future. It is not a crystal ball that will definitively find the exact threat picture 10 years from now. Instead, it provides a range of threats that you might not have been aware of in order to do something about it.

A good example of the power of this kind of thinking was expressed by General Dwight D. Eisenhower on November 14, 1957 in a speech to the National Executive Reserve Conference in Washington DC.

> *"**Plans are worthless, but planning is everything**. There is a very great distinction because when you are planning for an emergency you must start with this one thing: the very definition of 'emergency' is that it is unexpected, therefore it is not going to happen the way you are planning."*

President Eisenhower is highlighting the idea that planning puts people in a position for success. When events or threats unfold, people are that much closer to making the detailed decisions to ensure a successful outcome. The action of planning allows people to hone their minds intellectually and to be seeped into the character of the problem so that they can quickly make changes on the fly.

Why 10 Years in the Future?

The Threatcasting Method focuses on ten years in the future. This is a conscious timing decision in order to reduce innate biases from participant(s) when they participate in the Threatcasting Workshop and explore possible threat futures. The ten-year time horizon allows for, and overcomes, plausibility concerns. For most, envisioning ten years into the future is an intellectually freeing experience, allowing participant(s) to imagine a broader range of futures beyond their current state. Typically, the ten-year time horizon is freeing because it is past the duration of:

- political administrations;

- corporate executives' appointment;

- the life cycle of most projects; and

- the current career or life position of the participant(s).

Using the Threatcasting methodology, the participant(s) can free themselves from the baggage of their emotional and intellectual connections to the present, allowing them to envision the future.

The 10-year time horizon is also a helpful guide when curating the prompts that will be used to influence the participant(s). A decade is near enough that current research applies and not so far away that the prompts would be implausible.

Many organizations and businesses will plan for the future, but their planning is generally a few fiscal quarters out or possibly two to five years. These strategic plans focus on current conditions, incremental change and business metrics. The ten-year horizon can help these planners "leapfrog" their current planning activities, expanding the range of the possible and probable futures. This expansion could identify threats that had been previously missed.

Ten years is also important as this method develops many harsh visions of the future that could be full of death and destruction and/or a significant change to our way of life. But ten years is enough time to derail these futures. Participant(s) are empowered to develop solutions to ensure that these negative futures don't occur.

The Six Phases of the Threatcasting Method

The Threatcasting Method is broken into six steps or distinct phases that contain tasks and activities. These phases are meant to provide the analyst(s) structure and guidance to conduct the Threatcasting. They are rigid and need to be followed closely. A phase cannot be omitted or skipped. The tasks and activities inside of each phase need to be performed before the analyst(s) can move on to the next phase.

The phases are numbered zero through five. For some, the idea of starting with a Phase 0 (zero) might seem strange. Phase 0 is full of the preparatory actions before the Threatcasting Workshop and other activities begin. In the military, Phase 0 is defined as "shaping the environment" and contains activities designed to set the conditions for success in future phases. For technology development projects, Phase 0 is the scoping and preparation part of the project. Often this is seen as the most important phase as all the following phases build upon it.

Note: We will go into greater detail in the upcoming chapters of the book.

Phase 0: Preparation and Curation (Pre-Workshop)

The beginning phase of the method consists of the preparation of the project and the workshop. This phase also includes the curation of the team, participant(s), and research prompts.

The initial action for the analyst(s) is to develop the **Threatcasting Foundation,** consisting of:

- the topic area to be explored;

- the specific research question; and

- the area(s) where the findings will be applied.

Informed and guided by the foundation, the analyst(s) pulls together a team, determines who should participate in the workshop, and decides what research or inputs should be used as prompts to envision threats 10 years into the future. Finally, the materials (e.g., workbooks, presentations, support materials) are created to conduct the workshop.

Phase 1: Prompt Presentation, Research Synthesis, and Discussion (Workshop)

The next phase begins the actual Threatcasting Workshop. Analyst(s) use the prompts and materials to engage in a participatory design session with participant(s). This activity presents the prompt to the participant(s) and then leads them through a session to explore the ramifications of the prompts, capturing their discussion in workbooks for use later by both the participant(s) in the following stages of the workshop and by the analyst(s) in the post workshop phases.

Phase 2: Futurecasting (Workshop)

Guided by the prompts and research synthesis, participant(s) engage in another participatory design session to envision a possible and potential threat ten years in the future. Participant(s) move from the high-level macro view of the research synthesis and prompts to the micro perspective of a person in a place experiencing a threat. To do this, they follow the Science Fiction Prototyping (SFP) and Experience Design Processes to generate a qualitative Effect-Based Model (EBM).

Phase 3: Backcasting (Workshop)

Using the EBM, participant(s) begin backcasting in small groups, developing a time-phased, alternative-action definition (TAD) phase that generates specific actions that can be taken to disrupt, mitigate, and recover from the threat. Additionally, participant(s) identify the indicators (flags) over the next decade that will show that the threat is beginning to manifest and become a reality.

Phases 2 and 3 can be repeated multiple times during the workshop to generate a high volume of threat futures.

The research synthesis workbooks along with the Threatcasting workbooks make up the Threatcasting Method's raw data to be processed by the analyst(s).

Phase 4: Post Analysis, Synthesis, and Findings (Post Workshop)

After the conclusion of the workshop, the analyst(s) study the raw data, using multiple techniques to cluster and identify the possible and potential threats. These findings are documented and sometimes peer reviewed by the participant(s) and Subject Matter Experts (SMEs). Additional research is conducted if needed and the technical documentation captures the threats, actions, and indicators.

Phase 5: Output (Post Workshop)

The final phase of the method translates the findings into an output. This output is determined by the Threatcasting Foundation in Phase 0. The correct output (e.g., technical report, academic paper, podcast, etc.) is determined by the person or organization that will be applying or using the Threatcasting findings.

Applying the Method

The Threatcasting method is distinct from traditional notions of futures thinking, planning, and modeling. Not only does the methodology combine both linear and creative thinking, it also requires that a diverse set of participant(s) from multiple organizations and domain expertise gather and collaborate. This diversity of participant(s) and the multidisciplinary nature of the sources it draws upon paired with multiple guided exercises to explore possible threats enables groups to envision a complex and evolving threat landscape.

1.2.2   THREATCASTING AS A FRAMEWORK

"Frameworks are by definition a little loose. They exist to provide structure and direction on a preferred way to do something without being too detailed or rigid. In essence, frameworks provide guidelines. They are powerful because they provide guidance while being flexible enough to adapt to changing conditions or to be customized for your company while utilizing vetted approaches."—Scott Ellis (2008)

From Method to Framework

Once an analyst(s) understands the specific steps, tasks, methods, and, ultimately, outputs of the Threatcasting Method, they can begin to use Threatcasting as a Framework. Within the confines of the methodology, the analyst(s) can use Threatcasting as a framework to explore threat areas, problem sets, and timelines.

The framework is designed to provide flexibility and serve as a guideline. The framework is the outline of tasks to execute the Method. However there is some "wiggle room" for other tools or practices to be used in order to achieve the end goals of the Method.

Threatcasting Framework

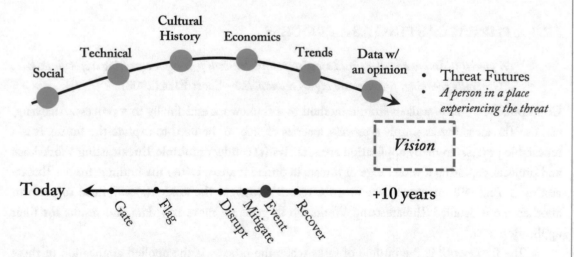

Figure 1.1: Threatcasting framework.

The Threatcasting Framework (Figure 1.1) captures the tasks of the methodology but allows the analyst(s) to make decisions to curate and hone the activity. This will depend on the Threatcasting Foundation, specifically the applications area(s).

Example Adjustments

The goal of the framework's flexibility is to allow the analyst(s) the ability to make changes and adjustments to work with the participant(s) to investigate a wider range of threat futures and provide detail in areas that relate to how the results will be used.

This understanding could modify the Threatcasting Foundation or the details for how the analyst(s) conducts the workshop.

Additionally, in the midst of the workshop the analyst(s) may need to adjust a specific task, prompt, workbook, or detail to generate the results and data needed for the application area(s).

Example Adjustments:

• Specify a threat actor

• Modifying the location

• Limiting the prompts

• Adding new information

The Method remains the same, but when used as a framework, Threatcasting can adjust and flex to accommodate the specific needs of the Foundation and the necessary generation of the raw data needed for the post analysis and findings.

1.2.3 THREATCASTING AS A PROCESS

> *"A process is simply a well-defined set of steps and decisions points for executing a specific task. …Generally speaking, processes are highly repeatable."*—Scott Ellis (2008)

Conceptually, we have walked from a method to a framework and finally to a process. Analyst(s) can use Threatcasting as simply a process, a series of steps to be used to explore the future. It is a repeatable process. In many application areas, analyst(s) conduct multiple Threatcasting Workshops and projects, exploring a wide range of threats in different areas. Often the findings from a Threatcasting project will uncover a new area for research wherein the analyst(s) will need to perform another, more detailed Threatcasting Workshop to generate more raw data and results for their application area.

The framework is the outline of tasks while the process is the applied application of these tasks to reach a specific result. The process could and usually is different depending upon the desired outcome and the application of the findings.

Part 2 of this book outlines the process for designing and executing multiple sizes and instantiations of Threatcasting. We provide a series of steps and decision points for the analyst(s) to follow when applying Threatcasting to a specific problem and organization.

1.3   FLIPPING THE SCRIPT ON CYBERATTACKS

Interview

Ron Green is the Chief Security Officer of Mastercard, a multinational financial services corporation. Under the guidance of Green, Mastercard has been using Threatcasting for years to explore potential threats to consumers and their business as well. Green shares his thoughts about the usefulness of Threatcasting.

The world we're in today looks very different than it a did a decade ago. How we shop, pay, interact, and connect has changed—we're living in a digital-first world. Our approach to security must change as well.

At Mastercard, we're working to build the world of tomorrow. But how do we make sure we're secure? How do we stay ahead of the threats that we'll see in the next decade and beyond?

Given the textbook you're currently reading, it should come as no surprise that a key element of our work in this space is Threatcasting. Which is exactly what it sounds like: threat forecasting.

Each year, we conduct our own Threatcasting exercise. We convene global subject matter experts across a wide variety of cultural, sociological, economic, and scientific fields, from across the public and private sector. Together, the group looks beyond "what's around the next corner" and envisions the worlds that might exist 10 years from now. By going beyond the 1–,3– and 5–year horizon, the team is better able to unlock their creativity, ingenuity, and innovative spirit.

The Threatcasting framework gives us the ability to combine the wide range of expert inputs through a series of information sharing exercises that help us imagine a broad range of future threats. It also giving us a systematic way to look backwards from these futures to understand the steps needed to disrupt, mitigate, and recover from them as well.

When we think about a topic as complex as security, it is important to note that we're not just thinking about one, singular, future. We're thinking multiple times about multiple futures involving different types of people around the world. From there, we're able to step back and ask, "what do we need to do as an organization, as a nation, or as an industry to prepare ourselves for those futures?"

Put simply, we're using all these expert inputs to build a roadmap for many potential futures. It shows us not only what might be coming, but also how we would need to respond. It is an important part of our ability to stay nimble as we combat oncoming threats and be deliberate in how we prepare for what's next.

Threatcasting has helped us organize our thinking around topics like Quantum, Supply Chain, and communication between ground and Space-based assets. We even hosted a special pandemic-focused session just as COVID-19 took hold of the globe in March 2020.

Through that Threatcasting session, we were able to understand how we could prepare for the road ahead and what we could do to minimize the disruption to our workforce, both in and out of our physical buildings. We left with a sense for where we should focus our efforts and where additional help was needed.

Blending collaboration and innovation in this way are essential elements of how we secure ourselves today and help us build a sustainable, resilient security posture for the future.

1.4   POETRY

Relativity
for Stephen Hawking

When we wake up brushed by panic in the dark
our pupils grope for the shape of things we know.[1]

Sarah Howe

Why This Poem Matters

"If we can think this far, might not our eyes adjust to the dark?" Howe's final verse captures a rally cry for Threatcasting. Her poem begins with a waking panic in the dark, our eyes search the room shooting around protons like greyhounds. But then she speaks to the wonders of physics and science to understand these sometimes confusing and complex concepts. Everyone who uses the Threatcasting Method uses science, technology, engineering, art, design, and research to better understand the future, to allow our eyes to adjust to the dark.

---

[1]   Read the full text at https://www.theparisreview.org/blog/2015/10/08/on-relativity/ or https://bit.ly/3kPlQHe.

CHAPTER 2

Phase 0

"The beginning is the most important part of the work."
—Plato

2.1 PREPARATION AND CURATION

In U.S. Department of Defense terminology, Phase 0 is a shaping phase—namely, influencing the state of affairs in peacetime before an operation. Shaping activities (occurring in this phase) help set the conditions for the successful execution of a military operation. Similarly, Phase 0 in the Threatcasting Method consists of preparation for the project as well as curation of the research that will be used and participant(s) that will be a part of the workshop. Careful preparation and curation will ensure the effectiveness and clarity of the results.

In technology product development, Phase 0 is often the first phase of a project. A Phase 0 converts the core product ideas into detailed specifications that can then be fully designed and implemented. Often during a Phase 0 project teams collaborate with clients, end users of the project, on the features, functions, and interfaces of a new product, generating a well-defined set of requirements. "Phase 0 Projects provide the structure...to define the problem and evaluate potential solutions before diving into the detailed design implementation" (AppliedLogix, n.d.).

The activities within this phase are:

- develop the threatcasting foundation;

- assemble the core team, steering committee, and analyst(s);

- select and gather research prompts;

- select the participant(s); and

- draft the threatcasting workshop workbooks.

2.1.1 THREATCASTING FOUNDATION

"To ask the right question is harder than to answer it."—Georg Cantor

Phase 0 begins with the analyst(s) developing the Threatcasting Foundation. The foundation (see Figure 2.1) has three main components: the topic, research question, and applications area along with an understanding of how the output of the Threatcasting Method will be used. Establishing a robust, researched, and focused foundation is an important skill for any research project. The better defined the foundation, the more efficient and effective the application of the Threatcasting Method will be.

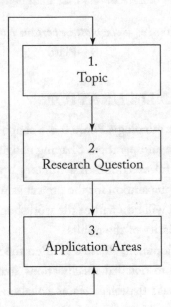

Figure 2.1: The Threatcasting Foundation is a three-component iterative process that provides the analyst(s) the premise for the entire threatcasting project.

Topic

The topic for the Threatcasting is the high-level subject or area that the analyst(s) want to investigate. The topic can be broad (e.g., climate change, the future of cyber crime, future healthcare threats). It is meant to give the analyst(s) a place to start their broad research and also be able to communicate the nature of the project to others (e.g., additional analyst(s), participant(s), core team, steering committee).

The topic can be determined in multiple ways depending upon the application of the methodology and the desired Application Areas. These Application Areas are the third component of the Threatcasting Foundation. Once the analyst(s) has decided upon their topic, research question, and applications areas they may need to return to the topic to make sure it sufficiently captures the nature of the project.

At its most simple, the Topic is the area or subject that the analyst(s) want to explore. However, a sponsor or third party who are identified in the Application Areas or who have assigned the topic (e.g., professor, corporate department, government agency) can provide the topic. Wherever the Topic comes from, it is the first building block in the Threatcasting Foundation.

Before moving from a topic to the research question, analyst(s) should gather background research, investigate other explorations of the topic, and explore broader related or ancillary topics. This early research around the topic will help analyst(s) later during curation of the prompts and the participant(s).

Research Question

Once the topic has been selected, the research question aims to narrow the focus. It is important not to try to "boil the ocean".[2] If the research question is too broad, unclear, or not specific enough, the Threatcasting Method will not yield the optimal results.

There are a simple set of questions that can help ground the research question. (George Mason University Writing Center, 2008).

Start by asking *explore* questions. Ask open-ended "how" and "why" questions about the general topic. Consider the "so what" of the topic. Why does this topic matter? Why should it matter to others? Reflect on the questions that have been considered. Identify one or two questions that are engaging and that could be explored further through research.

Next, determine and evaluate the research question. What aspect of the more general topic do you want the analyst(s) and participant(s) to explore? Is the research question clear? Is the research question focused? Additionally, a research question must be specific enough to be well covered in the resourcing (i.e., time, space) available. Is the research question appropriately complex? Questions shouldn't have a simple yes/no answer and should require research and analysis.

Sample Research Questions

Clarity

Unclear: Why are social networking sites harmful?

Clear: How are online users experiencing or addressing privacy issues on social networking sites like MySpace[3] and Facebook?

Focused

Unfocused: What is the effect on the environment from global warming?

Focused: How is glacial melting affecting penguins in Antarctica?

Simple vs. Complex

Too simple: How are doctors addressing diabetes in the U.S.?

---

2   "Boil the ocean" is an idiomatic phrase that means to undertake an impossible task given resources available. It is derived from the literal aspect of boiling the ocean, which is an impossible task given the size of Earth's waters. To avoid boiling the ocean, projects should be clearly scoped within the realm of the possible.

3   MySpace is an American social networking service that from 2005–2008 was the most visited in the world (surpassing even Google visits). In 2009 with the rise of Facebook, MySpace rebranded itself to focus on music.

Appropriately complex: What are common traits of those suffering from diabetes in America, and how can these commonalities be used to aid the medical community in prevention of the disease?

Application Areas

The Application areas specify how the output from the Threatcasting Method will be used and who will use it. Threatcasting isn't performed in a vacuum. As an applied methodology, the findings outline steps that can be taken by organizations and individuals to disrupt, mitigate, and recover from potential threats.

<u>The Who</u>

The analyst(s) will need to specify who will use the findings.

> Example Organizations:

- Academic Research

- Corporations and Private Industry

- Government Policy Makers

- Military or Security Professional

- Trade Associations and Non-Profits

- Communities and Community Organizers

Each of these organizations will have different requirements and needs to apply the results of the Threatcasting. Understanding who will apply the results will modify how the findings and recommendations are drafted.

<u>The How</u>

The analyst(s) will also need to know how that desired organization is planning on using the results. How it will be applied will modify many of the questions that are asked to generate the raw data and the language used to explore the threat futures with the participant(s).

> Consider the implications and areas the results might be applied.

- What might the results say?

- Why does the argument matter?

- How will the results be put to use?

- How might others challenge the argument?

- What kind of sources will be needed to support the argument?

- What are the budgetary, organizational, stakeholder, and social capital requirements to enable application of the results (think short, medium, and longer term)?

Review and Revise

Based upon the details and specifics of the Application Areas, the analyst(s) may need to modify or revise the topic and/or research question to meet the requirement for impact on a specific group or organization. It is typical to conduct multiple iterations of review on the Threatcasting Foundation to ensure it is solid and will yield the desired output. Recall the goal of the Threatcasting method is to be applied on hard problems, so continue the revisions until the Foundation is solid. The Who and the How of the application areas can give more detail or specifics to the topic and research question.

Interview

Brian David Johnson is the Futurist in Residence at Arizona State University (ASU), a Professor of Practice in the School for the Future of Innovation in Society and the director of the ASU Threatcasting Lab. Brian is the creator of the Threatcasting Method. Brian provides some advice on drafting a Threatcasting Foundation.

Finding the Right Question

Ask any scientist or researcher, and they will tell you that finding answers to questions isn't the hard part. If you get the right people together you can generally get the answer to a question. The really difficult thing is to ask the right question.

I tend to think of the topic as an aspirational area. The analyst(s) should ask, "What do I want to do? What areas do I want to explore? If I want to see future threats, where should I start looking?

That's the crux of the topic area... it's where the analyst starts. What follows is an investigation on the topic. How can you learn more? What has been said before and who said it? How can your work add to the body of knowledge?

As the analyst(s) researches, they might discover multiple "answers" to the topic. Certainly, they will find multiple perspectives and research.

Next, the analyst should ask, what is my specific question? What is the question I want to ask that has not been asked yet? This will get you to your research question.

The steps for crafting your research question are outlined earlier in this chapter. But I would like to echo the idea that the analyst needs to be specific: the more specific the question, the more defined the project, the better the outcome. A specific research question will help for all of the following phases of the Method.

The final part of the Threatcasting Foundation is the Application Areas. This is one of the facets of the Method that sets it apart and makes it unique from other futures modeling techniques and tools.

When finding the right Question isn't good enough

Threatcasting is an applied method. The output is meant to be used by an audience to accomplish a task. This is where simply asking or finding the right question isn't good enough. The analyst has to know who they are asking the question for and what that person or organization will do with the answer.

This adds a level of complexity to the Threatcasting Foundation that will guide the analyst through the phases but most importantly it will give guidance in the post-analysis phase. Post analysis can be difficult for the analyst(s) because there are times when they will be "drowning in data." The Threatcasting Method can generate hundreds of pages of data and in this phase it is up to the analyst(s) to make sense of it. The Threatcasting Foundation will provide clarity and focus for the analyst(s).

For this reason, it is important that the analyst(s) continually repeat the process of interrogating the topic, research question, and application areas while developing their Threatcasting Foundation. With a strong foundation, the entire structure of the Threatcasting work will be strong.

2.1.2   ASSEMBLE THE CORE TEAM, STEERING COMMITTEE, AND ANALYST(S)

With the Foundation for the Threatcasting Method in place, the next activity in Phase 0 is to assemble the applicable, broader team. The lead analyst should determine if a core team or steering committee is needed. Additionally, the lead analyst can determine if other analyst(s) are needed to collaborate through the application of the Threatcasting Method.

Designating a core team, steering committee, and other analyst(s) needed is required regardless of what form of the Threatcasting is being used (e.g., individual, small group, or large group). The details for each application of the process are explored in detail in Part 2. Generally, the larger the number of participant(s) who will take part in the Threatcasting Workshop, the more necessary a core team, steering committee, or additional analyst(s) become. However, even if an individual Threatcasting Method is being used, it can be helpful to utilize a core team or additional analyst(s) for collaboration, an outside opinion, and/or to validate results.

Core Team

The Core Team for a Threatcasting Workshop is derived from the Applications areas in the Foundation. Understanding who will use the result of the work can give an indication of who could be required to participate in the core team. Generally, the size of the core team is small, from three to five people. Core teams can be made up of people from multiple organizations, domains, and backgrounds. Ultimately, they will be the primary consumers and implementers of the final results.

Activities of the core team include:

- participate in periodic meetings to plan and prepare for the Threatcasting Workshop as well as review the findings;

- review and refine the Foundation;

- assist in the identification of Prompts and possible communication with SMEs if needed;

- participate in the Threatcasting Workshop;

- review and validate the Findings;

- socialize and Implement the Findings;

Steering Committee

A steering committee is generally used for a large group Threatcasting Workshop. However, they can also be used along with a core team for any form of Threatcasting. As the name suggests, the steering committee's primary purpose is to help steer the effort to success by providing insights. They do not make decisions—that is the core team's responsibility.

The steering committee is larger than the core team and is made up of a wider range of people. The steering committee has fewer activities than the core team but is helpful to broaden the reach and impact of the Threatcasting.

The steering committee can include people who will be the main consumers of the Threatcasting findings, but they can also include experts in the Topic or Research Question, secondary or tertiary consumers, as well as a broader set of people to socialize and publicize the results. Because the steering committee should be helpful (and not a distraction), it is important to choose its members carefully.

Activities for the steering committee include:

- review and comment on the foundation;

- recommend prompts;

- make connections to SMEs and participants as needed;

- attend Threatcasting Workshop (optional); and

- review and comment on findings.

Conversations in the Lab

A conversation between Cyndi Coon, Chief of Staff of the Threatcasting Lab and Natalie Vanatta, senior advisor to the Threatcasting Lab. They spoke about the steering committee concept.

Cyndi: It is important to have a steering committee. However, I think that often a core committee feels "Oh gosh, it's just one more hurdle." That it translates to more people to please and get to approve things. Yet, I personally feel like the best part of the steering committee is that we get diversity (in terms of gender, age, background, experience, education) that we wouldn't normally find. So, my pitch is always for having a steering committee.

Natalie: I concur. I think the most difficult thing for people to understand is those words "steering committee." They mean something totally different depending on which domain or which organization you currently work for. I think a lot of times the steering committee concept can be scary because people assume they will slow down the process and that it is about empowering a different group of people (other than the core team) to make decisions. Nope.

From a Threatcasting perspective, the steering committee is really just an extra set of eyes that can provide a little bit of oversight and give us some new ideas and really come to the table with a Rolodex of amazing humans that might be good for the project (either as SMEs or participants). They don't want to get into the "sausage making" that the core team handles. So, I think a successful steering committee is just another set of eyes that understand the process and truly understands what the final product needs to achieve and how to reach the expected audience.

Cyndi: Absolutely. That is why it is so important who you pick for the steering committee. Yes, you want a diverse group similar to the SMEs and participants. But you also want folks that are high level enough in the audience that you're doing the Threatcasting for and in that audience you want them to be able to inspire and empower to create change based on the results of the Threatcasting.

Natalie: Yes, and because they come at the problem from a different perspective, that's helpful to ensure that as we go through the process and craft research questions and curate participants, and ultimately figure out how to package the final results narrative, they will have some different insights and thoughts about how to do that and how to communicate it the most effective as possible.

Analyst(s)

Additional analyst(s) are generally used for small- and large-group Threatcasting Workshops. Bringing in additional analyst(s) can help with collaboration and support in the early phases of the

Threatcasting process, provide support during the Workshop and act as collaborators in the post analysis. (These are explored in more detail in Part 2.)

Selecting different analyst(s) to collaborate on the process allows for different perspectives and areas of expertise that can strengthen the findings.

Additional analyst(s) can be a part of the core team throughout the project, or they can just be brought in during the Post-Analysis portion of the project (Phase 4). When to involve the additional analyst(s) is at the discretion of the lead analyst to decide where they can be most effective.

2.1.3 SELECT AND GATHER RESEARCH PROMPTS

Prompts are used during the Threatcasting Workshop to inform, inspire, and direct the participant(s). Based upon the Topic and Research question, analyst(s) curate a set of prompts. These prompts can be pulled from a range of interdisciplinary areas outlined in the Threatcasting Framework. Prompts are represented as the blue circles on the top arc in Figure 1.1. These prompts will form the structural foundation of our visions of the future.

Typical prompt categories include:

- social science research;

- technical research;

- cultural history;

- economic projections;

- trends; and

- data with an opinion.

The prompt categories guide the analyst(s) to gather interdisciplinary perspectives on the topic and research question. Curate prompts from different domains to provide research, perspectives, data, and opinions about the future. This will create the framework for the "art of the possible" on topic areas and give guidance when participant(s) explore the Threatcasting research question.

Social Science Research

Start with a prompt that can help participant(s) understand how society could evolve over the next ten years in your topic area. This is because the future, at its heart, is about people. The prompt could capture ideas about social context or how humans will adapt or interact with each other in the future. This could also be focused on relationships and connections.

Technical Research

This prompt category can contribute to understanding an interesting technology that is relevant to the Threatcasting topic and research question. Typically, this is a technology that is currently in a Research and Development (R&D) phase within industry or has recently achieved a basic scientific breakthrough. Selecting an early-stage technology allows that technology to be coming into the market and, potentially, enabling a mass effect in ten years—thereby, having a significant influence within the Threatcasting Foundation.

The intersection of these two prompt categories (Social and Technical) will enable an exploration in how this technology will actually be used by future humans. Technology with a long R&D phase rarely is used in the same way as the initial engineers/creators believe it will be used.

Cultural History

History is an important prompt for the future. History is the language that people use to talk about and understand the future. Cultural history prompts are used to understand the context for the society or organization that might be affected by the topic or research question.

Additionally, a cultural history prompt can be used to illuminate the impact that the Threatcasting findings might have on the intended application areas. In other words, what is the cultural history of the people and organization that might implement the findings from the Threatcasting? Having an understanding of these populations can be helpful when designating possible actions and outcomes from the post analysis.

Economic

The Economic prompt explores the possible future economic decisions that might affect the people or locations touched by the topic or research question. These could be trends at either the macro or micro level depending on the Threatcasting subject. This is important to developing our understanding of the future because economies and their shifting nature will greatly influence how we live and what we can afford. This now builds on our future construct because we are thinking about how individuals are evolving, societies are evolving, technology is evolving, but also how our economy is evolving. This then builds into the way of life that people want and the way of life that people can afford. The difference between the two could be an interesting area to explore.

Trends

The trend prompt attempts to capture macro or micro trends that could affect the topic or research question. Generally, these trends are regulatory or cultural in nature. For example, governmental policy decisions that affect the topic. Cultural shift or public sentiment can be effective trend prompts. For example, how are the public and the media views of the future of artificial intelligence (AI) and its effect on the future of labor are changing?

Data with an Opinion

The Data with an Opinion prompt encapsulates a core idea that drives the Threatcasting Method. People build the future. The future is built by organizations. This prompt embraces the biases and opinions of people who are a part of organizations that will have a meaningful effect on the future threat areas that are being modeled.

If a large auto manufacturing company has an opinion about the future of cars and mobility, the prompt can be useful. Even if that opinion is not 100% correct, the prompt is useful because it is an indicator for the actions that this influential organization will take. The actions could have an influence on the topic area.

It can also be insightful to find two prompts in this section, especially if the prompts do not agree. Generally, people and organizations have conflicting ideas for what the future might be. These two opposing views give a broader range of possible and potential futures. This conflicting prompt for the Threatcasting will give a wider range of results. Typically, with opinion-based prompts, neither is completely correct or false. The answer usually lies in the middle between the two.

Forms and Number of Prompts

Prompts can take many different forms. The form ultimately doesn't matter as long as the prompt serves its purpose. The curation of the prompts across prompt areas when collected together and presented to the participant(s) during the Threatcasting Workshop should give enough data, research, guidance, and inspiration to explore a range of possible and probable threats that address the topic and research question.

Because the prompts are used during the workshop, they will need to be concise so that the participant(s) can understand them quickly and begin to process them. This processing might be simply applying it to the Threatcasting Foundation—exploring the effect the prompt might have on the future. Or the prompts serve as a catalyst for discussion between multiple participants(s). This processing and conversation is captured in the RSW and used to develop the EBM.

There is an art to curating the prompts. It gives the analyst(s) a chance to put their personal stamp on the project. The prompts need to be short but not too simple. They need to be in depth and thought provoking but not so voluminous that they take too much time to comprehend. They should inspire and challenge the participant(s). They should inform participant(s) about emerging technical, social, or trends areas so that the participant(s) are free to explore the threat futures that answer the research question.

The previous category listing of prompts is not exhaustive. If the prompt, in the collective, informs the participant(s) to answer the research question, then it is acceptable.

Likewise, the number of prompts is up to the curation and discretion of the analyst(s). The number should also be dictated by the goal of the prompts. Generally, analyst(s) select between four and six prompts to use during the Threatcasting Workshop. This range has been effective because it informs the participant(s) but doesn't overwhelm them with data.

Types of Prompts

- Research papers and articles

- Business Reports and Projections

- Keynote videos and TED talks

- Interviews

- Subject Matter Experts (SME) Video

- Analyst(s) primary research on the topic or research question

The following are some examples of good prompts and ones that may not be as useful for a workshop.

Bad Prompts

Typically, bad prompts are too long in length or too complex for participant(s) to comprehend and start working with them in the workshop. Additionally, bad prompts can be too vague, short, or unclear as to the implications for the research.

- 25-page research paper

- Complete book

- 45-minute documentary

- A short bullet pointed list in Powerpoint with vague statements (e.g., the future of money, cashless society, blockchain)

Good Prompts

Good prompts have enough depth to give the participant(s) something to think about and can be understood in a short period of time. The ideas or research behind these prompts can and often should be complex and subtle but they need to be presented in a way that is functional for the workshop. (Kind of like a short poem!)

- The analyst(s)' summary of a 25-page research paper in a few paragraphs

- 10-minute video about a specific subject

- A technology demonstration by an analyst(s) or an expert that quickly explains the nature of the technology and what it can do

- A short Powerpoint presentation by an expert

- Analyst(s) Powerpoint summaries of a specific research topic, often pulled from the topic research at the beginning of the project

- Short- to medium-length article(s)

- Subject Matter Expert Video

Prompts as Categories: "Buckets" or "Containers"

In some instances, the analyst(s) may need more than one piece of research for the prompt to effectively capture the category. An analyst can use a combination of research inputs for a single prompt. Meaning, the analyst(s) might use a video and a research paper to adequately capture or cover the category.

It can be helpful to think of the prompt categories as a "bucket" or "container." The analyst(s) can put one piece of research (e.g., a video) in that container or if needed they can put two pieces of research in the container (e.g., a video and an article).

This gives the analyst(s) more freedom and creativity to specifically capture the category in the prompt for the participants during the workshop. It is important not to overload the participant(s) with too much information in a single prompt. They need to be able to comprehend it quickly and begin using it.

Examples:

The analyst(s) wants to use information about Millennials' outlook on finances as a social prompt. To capture the nuances, the analyst(s) decides to use a video of a social science researcher discussing their work interviewing millennials about their financial goals. But this is also paired with a financial article that provides survey data of the saving and spending habits of millennials to round out the category.

- The analyst(s) wants to include information about the use of AI in hiring software as a technical prompt. The analyst(s) provides the participants a vision video from an AI HR service provider talking about future capabilities paired with an executive summary of an academic research paper on the perils of data bias in current hiring software.

Each of these examples shows how the analyst(s), when needed, can use multiple pieces of research in a single prompt.

Prompts used as Analyst(s) Inputs

Not all of the prompts need to be used during the Threatcasting Workshop. As a part of the research for the Threatcasting Foundation in the specific topic area, these prompts can be used to inform the post analysis after the workshop. Therefore, these prompts become a part of the raw data that is analyzed.

Subject Matter Experts (SME) Video Interviews

Subject Matter Experts (SMEs) video interviews are a type of prompt that can be a very effective tool for the analyst(s) when curating a Threatcasting Workshop. These videos or interviews differ from other prompts in that they are recorded or created by the analyst(s), core team, and/or the steering committee. These interviews are with a current SME on a subject that the analyst(s) would like to use as a prompt.

SME interviews can be effective because they are up to the moment, current research on a specific topic. These interviews can be tailored by the analyst(s) depending on what questions are asked and what topics the analyst(s) requests the SME to cover.

However, these videos do take extra effort and time on the part of the analyst(s), core team, and/or steering committee. The following are some practical tips for the analyst(s) when considering a SME for a Threatcasting prompt.

Finding an SME

SMEs are thought leaders that have opinions which the community respects. An effective way to find SMEs is to look at keynote conferences within the domain. There are various websites that highlight the top 10 conferences in a certain domain. Go to those conference websites and take a look at the keynote speakers over the last three to four years. Another idea is to search the TED talk repository for individuals that have spoken on the topic you are looking for.

The key is not to initially constrain the search by only focusing on people that the analyst(s) or their core team and steering committee can contact. It is important to get thought leaders within the domain as they will be setting the structural foundation of the work and therefore will help ensure the work is grounded in science fact.

The Interview

Once the SME has been identified, it is important to understand what the analyst(s) will be asking them in the interview and what you will be asking them to do. The SME will not be a part of the Threatcasting Workshop. This is a very deliberate choice because the SME is a recognized expert on a subject and if they participated then the participant(s) in the workshop might defer to the SME, thus not gathering all the raw data needed for the post analysis.

The analyst(s) will interview the SME or ask them to record a 5–10-minute video clip where they will explore their area of expertise as a prompt directly to the Threatcasting participant(s). The video quality does not need to be high. In fact, it is often optimal if the video doesn't look produced or too flashy. The participant(s) in the workshop will feel they are getting a window into the SME's current research as it relates specifically to the curated area that will be explored.

Interview

Steven Latino is a facility Security Officer at ASU Research Enterprise (ASURE) and a U.S. Army Veteran. Steve handles the Threatcasting Lab's classified projects. He shares his ideas on selecting and interviewing SMEs for the Threatcasting Lab and what attributes to look for.

There's a quote in the military: "Even the best laid plan does not survive first contact with the enemy." Meaning, I might have the perfect plan, but once bullets are flying the plan has got to change. That's really the attitude you have to take when you ask SMEs leading questions and the response they give are often so unexpected, so left field. It's amazing for Threatcasting prompts. It's not necessarily about the question I'm asking. It's about getting the SME to freely discuss the topic and be very vocal in certain directions that as soon as we ask them a question, they fire off and they keep going on.

Before we hit record, we explain what it is that Threatcasting is exploring, who we are and what type of people we are, and maybe even explain how previous Threatcastings have gone to allow this person to set the stage in their mind, leading the conversation. And the questions that we asked prior to hitting record do a good job of getting SMEs to tell us who they are. When they tell us who they are, we can start pulling out nuggets of information we couldn't find on their bio that really steers the conversation the rest of the way.

And that's the whole point of the Threatcasting: to be in the space that we're uncomfortable thinking about. As long as we're operating in that space, we're good. So we have to get the SMEs talking about that uncomfortable space so the participants will live in that uncomfortableness, allowing the good to come out of it; so that the planning and the working toward a better future comes out of it (the workshop).

.... when we're talking about Threatcasting, we're really talking about multi-domain and multi-discipline logic. We're talking about a series that goes across multiple types of disciplines. I think one of my favorite things about Threatcasting as a participant is that every time there's been SMEs, if we were talking about something like drones, they weren't all drone engineers. There were also social scientists, data scientists, and other disciplines and domains. Government, commercial, and academics contribute to this broader base of knowledge so that we're not hyperfocused on whether the machinery is going to fail. We're also focused on the socioeconomic impact of that machinery failing. And we're focused on this grand scheme. That to me is amazing and it makes the Threatcasting an amazing tool.

2.1.4   SELECT THE PARTICIPANT(S)

"If everyone is thinking alike, then somebody isn't thinking." —General George Patton

The application of the Threatcasting Method can be done as an individual or a group. The differences will be covered extensively in Part 2. Regardless, the importance of people remains a cornerstone of the Threatcasting Method.

As important as it is to select a clear topic and research question for the Threatcasting effort, selecting the correct participant(s) for the Workshop is also essential. Participant(s) should be a diverse group of individuals—diversity of thought, experience, and expertise. Part of the group should be thinkers and part of the group doers.

As Threatcasting is a theoretical exercise undertaken by practitioners, it is vital that the majority have domain knowledge of how to specifically disrupt, mitigate, and recover from the theoretical threat futures. However, a few participants are curated to be outliers, trained foresight professionals, and young participants for fresh and multi-generational perspectives.

When Threatcasting on public sector topics, a good mixture of participant(s) should come from academia, private industry, government, military, non-profits, and think tanks. When using Threatcasting on corporate topics, participants can come from every aspect of the business, all levels and wide range of tenure within the company. You might also want to consider other partners in a corporate ecosystem. For example, Coca Cola might benefit by having a distributor or bottler in the mix. Working to step through potential intricate company politics can make the participant selection interesting. For instance, Target sells Coke. Target buys it from Coke. Having a Coke salesperson and a Target buyer in the room could be good (they share a reference frame) or bad (they negotiate against each other).

Diversity

Diversity can't be stressed enough as an important underlying value in the Preparation and Curation phase. This is referring to a diversity of ideas that can hopefully lead to a conflict of ideas. Every organization, whether a "mom and pop" business or a one-million person company, tends to struggle when thinking about the future as they have ingrained cultural ideas and unit identity that stovepipes their thinking.

Conflict

It is essential to the Threatcasting Method to bring together people that approach problems and think radically different from each other (either by training or experience). These participant(s) can have a conflict of ideas. From this conflict, participant(s) can reach new and novel solutions to future problems and threats. No one organization or individual has the answers, but by working together, participant(s) have the opportunity to succeed.

Bias

Bias is an issue that will come up throughout the Threatcasting Method. There are many different kinds of bias (Desjardins, 2017) (e.g., cognitive (Tversky and Kahneman, 1982), implicit (Perception Institute, n.d.), unconscious, cultural, etc.). Entire studies of bias have been done in the fields of research, foresight, participatory design, and many more (Meissner and Wulf, 2013; Bradfield, 2008). Bias is too large and important of a topic to cover lightly in this textbook.

However, the Threatcasting method does not shy away from bias. Quite the opposite it embraces it, accepting and acknowledging bias in all its forms and its potential effects on the interaction and output of the Threatcasting Method.

Analyst(s) should understand their bias and the bias of others on the core team, steering committee, participants, as well as the additional analysts. By confronting bias, recognizing it, and actively working to counter the effect, the analyst(s) can work to address the effects of bias.

When curating groups, research, and additional analyst(s), the lead analyst(s) should understand the existing bias (Project Implicit, n.d.), how it might affect the results, and what other additions might be needed to make the output of the Threatcasting Method as robust as possible.

Working Groups (where applicable)

During the workshop, the participant(s) will break into smaller working groups (typically three to four people). Chapters in Part 2 provide an in-depth discussion on how to create these small groups in order to maximize creativity and output.

The placement of the participant(s) into these groups is an essential step in Phase 0. The makeup of the teams can have a serious effect not only on the threat futures that are modeled, but also on the quality of the raw data. Further discussion and specific actions are outlined in Part 2.

These working groups will remain the same throughout the Threatcasting Workshop. The first coming together of the working group allows the participant(s) to get to know each other and get comfortable collaborating.

This social aspect of the Workshop is important but not a primary function of the Threatcasting Method. However, several participant(s) have called out these small working groups—the socialization and the discussions—as being an important benefit for participating in the workshop. They meet people that they would, otherwise, not have had the opportunity to ever interact with. These expanded networks enable them to create new solutions and think differently about their ecosystem once they return to their parent organization.

In recent years, Threatcasting Workshops have connected researchers from different universities who began collaborative research projects. The events have connected private industry (e.g., Silicon Valley, financial industry) with government agencies and national defense agencies. Many of these collaborations grew out of the realization that many future threats cannot be addressed by a single organization. These "whole of industry" or even "whole of nation" problems demand wider collaboration and communication that began at the Threatcasting Workshops.

Conversations in the Lab

The conversation continues between Cyndi and Natalie. In this segment, they are discussing what to look for in participants.

Natalie: I think there are two important things when it comes to picking participants. First, they need to have an open mind. And second is a diversity of thought and experiences, which generally comes from a diversity of gender, age, and background. Because when we curate a diversity of thought, and we put folks into a small group, they will push each other to go past what seems normal for each other. If you make a small group with three or four or five people that have experienced life differently and that have different domain expertise, you are going to get a much richer and fuller vision of what the future might look like.

Cyndi: I would add two other things that are vital for participants to have. Imagination and creativity are super vital to the process. I think it gives a permission space for that diversity of thought so you know how to handle the inputs. If there is fear around being creative, then it negates some of the process. It's the best moment when we're working on some crazy threat and a team starts building a movie or something like that, so that they can see through a lens (because they find out they're all visual). That's a cool moment. It is important for the participants to be able to grow their ideas through their imagination, through their creativity, through these unique lenses that each of them is bringing.

Natalie: Absolutely. And finding that right fit of humans for the room is not an easy problem. We have to start out as planners with a high-level understanding of what we want. For instance, a certain percentage to come from an academic background and a certain percentage from a government background. Then overlaid with the next layer of what we want—maybe a certain percentage from the tech industry versus a social science perspective versus a different perspective. In the end, we develop an algorithm of sorts focused on diversity characteristics that we believe will generate the diversity of thought within the room. And then we just start throwing names against those output descriptors of participants.

Other Individuals

For some Threatcasting Workshops, a facilitator could be needed. This is typically for a small- or large-group session (see Part 2). A facilitator leads the participant(s) through each step of the threatcasting process, answering questions, giving guidance, and helping to shape or inform the participant(s) threat futures. Facilitators are not participant(s). They have a separate set of tasks. Often, the lead analyst will fill this role but that is not a requirement.

Along with Facilitators there are other non-participant roles in a Threatcasting Workshop. Co-Facilitators are people who do not speak at the front of the room, but do help answer questions during the small working group work sessions.

Trained participant(s) are people who have been trained or who are experienced in the Threatcasting Method. They typically are assigned one per small working group to help lead and answer questions. However, trained participants are participant(s) and are meant to collaborate with the others in the group.

More information about Facilitators and how they apply to the large- and small-group workshops can be found in Part 2.

2.1.5   DRAFT THE WORKBOOKS

The workbooks are the main tool used to capture the data that will be needed for the post analysis. The workbooks give the participant(s) a place to write down their thoughts and perspectives on the prompts as well as a place to capture the EBMs, which are the main output of the Threatcasting Workshop. There are two types of workbooks used in the Threatcasting Method: Research Synthesis Workbook and the Threatcasting Workbooks.

These workbooks can be a simple structured data gathering tool or a heavily designed (e.g., language, visual design) tool.

Technically, the workbooks are a structured database (e.g., spreadsheet, online forms, etc.) where the information is captured. They are then used by the analyst(s) in the post analysis or as an input to other data processing platforms. Details describing how these workbooks are used and analyzed are in Chapters 3, 4, and 5. They are touched upon here in Phase 0 because it is a preparation activity to create them. However, it is easier for the reader to understand the reasoning behind their design if we showcase them in upcoming chapters as they are used in the process.

Research Synthesis Workbook (RSW)

The RSW is used with the prompts to capture the participant(s) analysis of the curated inputs. The RSW uses a series of open-ended questions to draw out the perspectives and opinions of the participant(s). This workbook is described in detail in Chapter 3.

Threatcasting Workbook (TCW)

The TCW contains a series of questions and tasks that leads the participant(s) through the design and modeling process to develop their EBM. Using the tools of Experience Design, the participant(s) explore a person, in a place, experiencing a threat—all based upon the prompts and the RSW information. This workbook is described in detail in Chapters 4 and 5.

Interview

Renny Gleeson is the founder of the Business Innovation Group (BIG). Prior to this, he spent over a decade at Wieden+Kennedy, the world's largest independent advertising agency, as the Global Director of Interactive Strategy and the Managing Director of the Business Strategy Group. We asked Renny to tell us how he approached the creation of the workbooks, turning them from spreadsheets into art for a Threatcasting that he was involved in.

First things first...

I've worked in advertising and digital marketing for more than twenty years. An ad agency's job is to get people to buy their clients' products. To do that, we inform, we entertain, and we create desire—and we do that with stories. Some of those stories are shaped like ads, some aren't, but when there's a range of products equally capable of meeting the customer need your product solves, you don't win that sale with logic. The better story wins. Because people don't just buy a product, they buy the story about themselves that product helps them tell, and they pay a premium for stories about the better, sexier, funnier, more compelling person they'll become with that product in their life. Once our most basic needs are met, we don't buy products. We buy stories.

Help People Tell the Best Stories

That's why Threatcasting is so interesting: stories are at the core of the EBM worksheets. And while a great story can be hard, the Threatcasting Method walks people who might not consider themselves "storytellers" step-by-step through a creative ideation process—and some pretty damn good stories come out. Some scary ones, too. But then that's the point.

My business strategy team worked with our awesome design group to craft workbooks for an exercise exploring the future of AI in a home environment. I worked with the team to determine each question in the EMB and how best we could use it to inspire good thinking and even better stories. The experience and enabling questions weren't just there to interrogate people about potential threat futures; each question gave space and oxygen to their creative process and drew out new aspects of their story. We chose questions that drove reappraisal and unlocked "aha" moments. The books spurred them to think broadly and granularly, then rethink again, and the result was twists and turns that surprised them even as they were writing them... and made their stories so much better.

Questions are reminders for places to look. Helpful blazes on an unknown trail. We ask ourselves: what experience questions might inspire unexpected insight? For example, "what does the person experiencing the threat smell when the event happens" or "what do they eat for lunch?" or "where might the person in our story go to cry?" These are questions out of left field that kick in the empathy, that remind them they aren't just telling "a" story, they're telling a real person's story. Someone they might have more in common with than they thought.

The same thing is true for the enabling questions: how often do you get to put on your bad guy hat? All bets are off when you enable the bad guy or gal or group. Every villain is the hero in their own story, so how do you make the best villain possible?

There's a kind of game going on behind the scenes in the workbooks. The better and more interesting the questions you ask, the better the answers you get, and the better the raw data is for post analysis. Getting people to tell great stories is fun, but it's also productive. It makes for great results.

We Also Make the Best-Looking EBMs

Did I mention we were in advertising? We reimagined the basic EBM spreadsheets to create a desirable object. We created note pads wrapped in designs our workshop participants could take home (Figures 2.2 and 2.3). And they did. We wanted the experience of going through a Threatcasting Workshop to feel exclusive and a little dangerous. We wanted it to feel cool and sci-fi and like you were actually writing down a dark future that was going to happen and it was going to be amazing and terrifying. I briefed the designers to imagine a sci-fi villain's notebook. That became the branded booklet teams used to capture their stories and their data.

Figure 2.2: Front cover of the Threatcasting workbook.

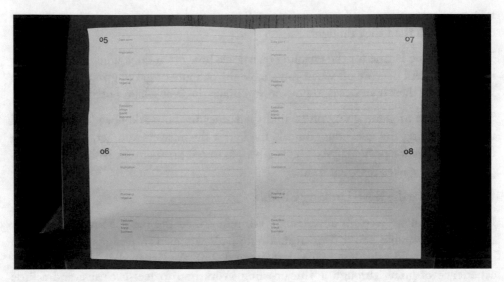

Figure 2.3: Inside of the Threatcasting Workbook.

There's a story behind the scenes when you build your EBM worksheets. They can be "forms to fill out" or they can "plus up" the experience. What's captured in those pages will drive your results, so don't miss the opportunity to inspire the best possible stories!

SPECIAL NOTE: More than Methodological Preparation

This phase outlines the steps that need to be taken to prepare for the next phase and the Threatcasting Workshop for the methodology. Depending on the type of workshop there may be more logistics that need to be taken care of. These are covered in depth in Part 2 as they all depend on which instantiation of Threatcasting you wish to perform.

2.2   POETRY

Innocence

The height of wisdom seems to me
That of a child; [4]

Robert William Service

---

[4]   Read the full text at https://bit.ly/39QNj4S.

Why The Poem "Innocence" Matters from Nikhil Dave

Nikhil Dave is a student at Arizona State University who conducted a Threatcasting Workshop in 2020 on Biological Big Data.

I have a firm belief that naivety is what allows for creativity, imagination, and novelty. Thus, I used the poem "Innocence" to start the four-day Threatcasting Workshop and to encourage participants to not fear naivety. I stated that there is certainly a portion of the topic that each participant would be naive about, regardless of their background/experience, and thus they must embrace that naivety to be creative and imaginative in the work they will do in the workshop.

2.3   EXERCISES

2.1

A topic selection is the kickoff inspiration that launches the research inquiry. Once the topic is selected, it is time to refine the topic into a Research Question.

Action

Create a research question using your topic as a lead off.

Steps

1. Pick a general topic.

2. Start your project documentation. Take a moment to decide how best you can capture your thoughts, ideas, and work that is most comfortable for you. This could be a text document or spreadsheet or post-it-notes or a combination of them all. Additionally, think about how you might want to share your work products with collaborators—remote-friendly tools (OneNote, GoogleDocs, etc.) might be appropriate.

3. Gather background research on the topic, including sources. You should have at least four sources to help provide a well-rounded literature review.

4. Write down ideas to explore nearby topics to the topic you selected. Include topics that are broadly related and ancillary related topics.

   In your project documentation, note the above research. Now, use your research to narrow the focus of your topic.

5. Draft your research question. Ensure that it is clear, focused, and appropriately complex.

2.2

Using the Threatcasting Foundation that was developed in Exercise 2.1, answer these questions.

1. Considering that the Core Team will need to attend meetings, help prepare for the workshop, assist in finding prompts, SMEs and participants, review and socialize the findings:

   a. What kinds of people would be helpful to have on your core team?

   b. What domains might they come from? (e.g., private industry, academia, government, military, non-profit)

   c. Would a mix of members make the Core Team stronger? Why?

   d. What expertise does the Core Team bring to selecting the prompts?

   e. Is there anyone on your Core Team who might be better on the Steering Committee?

2. Considering that the steering committee will need to make recommendations and connections to SMEs and participants, possibly attend the workshop, and review the findings:

   a. Do you need a steering committee? Why?

   b. What kinds of people would be helpful to have on your steering committee?

   c. What domains might they come from? (e.g., private industry, academia, government, military, non-profit)

   d. Would a mix of members make the group stronger? Why?

   e. What networks and resources would the steering committee need to share with the analyst?

3. Considering the analyst(s) will be conducting the post analysis and working with the lead analyst to possibly curate the workshop:

   a. What perspectives would be beneficial from additional analyst(s)?

   b. Do specific analyst(s) have an expertise that would add or augment the lead analyst?

   c. Are there conflicting opinions that might be helpful?

    d. Are there different areas of study or professional experience that an additional analyst could bring?

2.3

Continuing the work on Exercises 2.1 and 2.2, this exercise will help you think about selecting your subject matter experts for prompts.

1. Pick four general prompt categories that will have bearing on your research question. For each:

    a. Research names of thought leaders who have taught, published, or worked in the field of your selected topic.

    b. Write down at least three names, their affiliations, and area of expertise.

    c. Brainstorm a way that you might be able to approach the three individuals identified. Think of "Six Degrees of Kevin Bacon"[5] of how you might be able to connect to these individuals.

2. Reviewing your work to-date, is there a need for a fifth prompt? If so, what would it be?

Tip:

Wondering where to look for SMEs? Start by asking your network who might get you one degree of separation closer to connecting. Try keynotes listed for conferences, websites, papers, articles, books, and other lecture series such as TED talks for individuals that have spoken or written on the topic that you are looking for.

2.4

This exercise is designed to develop an ability to think more broadly about who is "in the room" during a Threatcasting Workshop. Curating a room of Threatcasting participants is an art form that can be mastered through practice. Let's practice building the room where Threatcasting happens.

1. Write down a list of names of individuals who work in the field of the topic you are focused on.

---

[5] Six Degrees of Kevin Bacon is a game where players arbitrarily choose an actor and then connect them to another actor via a film that both actors have appeared in together, repeating this process to find the shortest path that leads to American actor Kevin Bacon. The game's name is a reference to "six degrees of separation," a concept which posits that any two people on Earth are six or fewer acquaintance links apart.

2. Conduct your research. Note where people work, what projects and outcomes they have engaged in, other people they have worked with or co-authored with. Check that you also have people who understand the application areas in your Threatcasting foundation, also known as people who know how the final output will be used.

3. Build a spreadsheet with columns noting the following areas: domain (government, military, industry and higher ed), gender, generation, background, and links to bios. Filling out these criteria will allow you to see right away if you have a gap.

4. For every name you include see if you can find someone who might have a different perspective. Add them to the list.

5. Compile a list of at least six individuals that you think would be great, diverse participants to explore the Threatcasting Foundation.

CHAPTER 3

Phase 1

"Research is seeing what everybody else has seen and thinking what nobody else has thought."
— Albert Szent-Györgyi, Nobel Prize in Physiology or Medicine in 1937

3.1 RESEARCH SYNTHESIS

Phase 1 of the Threatcasting Method (Research Synthesis) begins with convening the Threatcasting Workshop. The workshop is the culmination of the planning and curation in Phase 0. This workshop could consist of a large or small group of people, or it could be an individual. We will discuss the specifics for each of these applications in Part 2. A workshop typically begins with reviewing any logistics and presenting the Threatcasting Method to the participant(s) so that they have an idea of how the workshop will be structured.

The activities within this phase are:

- present the curated prompts;

- complete the Research Synthesis Workbook (RSW);

- generate discussion; and

- prepare the participant(s) for Phase 2.

3.1.1 PRESENTATION OF PROMPTS

Workshop Note: Before the presentation of the prompts, discuss the goals and expected outcomes from the review of the prompts as well as the following Research Synthesis Exercise. (See further discussion in Part 2 in the Workshop Logistics for the Large and Small Groups Chapters.)

Presenting the prompts to the participant(s) will help define the future landscape while at the same time limiting the amount of information that the participant(s) have to work with. As the participant(s) continue through the workshop they will only use the prompt provided by the analyst(s) and their own knowledge as inputs to their threat futures. The participant(s) will bring their own perspectives as they envision future threats. This is a positive feature of the process as we want to capture the perspectives and opinions of the participant(s); the curated prompts give a place to start and can also inform or illuminate some participant(s) about new technologies, perspectives, and research or provide further illumination.

This phase draws its analytical strength from the Delphi method, which engages expert opinion through iterative rounds of questioning and feedback. The goal of the Delphi method is to converge on agreement after successive rounds of communication (Landeta, 2006; Linstone and Turoff, 2011).

"The Delphi survey is a group communication technique based on an interactive, sequential, and multi-step characterization of expert stakeholders, their interests, and intersection of interests. It has the advantage of obtaining opinion from experts, with a guarantee of anonymity, avoiding the potential distortion caused by peer pressure in group situations such as focus group analysis.

The classic Delphi study has three rounds: (1) a general questionnaire asking panel members to identify the pressing issues in a given knowledge domain (e.g., an emerging technology such as metagenomics); (2) a second-round questionnaire asking panel members to rate the importance of the list of the issues identified from the first round; (3) a third-round questionnaire, asking panel members to re-evaluate their ratings of each survey item after reviewing the expert panel's collective stance in the second round in response to the survey questions" (Birko, Dove, and Özdemir, 2015; Rowe and Wright, 1999; Beech, 1997; Delbecq, Van de Ven, and Gustafson, 1975).

The Threatcasting Method modifies the Delphi concept; it is adapted as the prompts provide research, data, and expert opinions on the current state of their domain and how it might evolve over the next decade. However, the iteration of these ideas come from the participants instead. Unlike Delphi, the goal is not a consensus on one idea, but to provide the unique data points that become the scenario framework and are, ultimately, integrated with participant insight and synthesis.

3.1.2   DATA CAPTURE: RESEARCH SYNTHESIS WORKBOOK (RSW)

The primary goal of the Research Synthesis phase is to "capture the wisdom of the room." The participant(s) will all have a personal perspective on the topics covered in the prompts.

The goal is to review the prompts and capture the participant(s) understanding, opinion, and perspective of them. The implications from the prompts and the potential high-level actions that could be taken in response to the prompts are some of the concepts that we want to capture from the participant(s). When captured in the workbook, this information can provide new perspectives and give counter arguments. It also serves as a "warm up" for the participant(s). The participant(s) begin to get familiar with the threat space in which they will be working in the following phases.

The questions in the RSW are specifically picked to both focus the participant(s)' attention on the threat space but also to give room to capture a broader range of data and information about the prompts. This data, captured in the workbook, will be essential not only for the following phases of the workshop but also for the post analysis and generation of findings.

Research Synthesis Questions:

Please note that these are the general questions typically used upon the participant(s) digesting a prompt. The specific wording is important depending upon the participant(s). The following are the typical RSW questions (also visualized in Figure 3.1).

- What was a data point that you found important or interesting?

- What are the implications of the data point on the threat futures or research questions?

- Are these implications positive or negative?

- What should we do about it?

There are no correct answers to these four questions. Each of these questions is purposely designed to be open-ended. "Open-ended questions are used alone or in combination with other interviewing techniques to explore topics in depth, to understand processes, and to identify potential causes of observed correlations. Open-ended questions may produce lists, short answers, or lengthy narratives..." (Weller et al., 2018).

Each question is used to urge the participant(s) to think about the prompt from different perspectives so that these perspectives can be captured in the workbook for later use. As the participants move through the four questions, they progressively get more open ended.

Data Point #	Summary of the Data Point	Implication	Is the implication Positive or Negative? Why?	What should we do?
1	consumers demand more transparency on where their food comes from	farmers will need to develop sophisticated tracking mechanisms on each head of lettuce from its lifecycle until it departs their custody	Negative - very expensive to do; will need more technology and connectivity at the farm	More research into Smart Farming technology (vice just assuming that IoT and other tech will work in farm conditions when it was designed for city use); Will need subsidies to fund at small farms; need to develop standards for the transparency at all phases of the supply chain (from field to table);
2				
3				
4				
5				
6				

Figure 3.1: Example RSW.

What was a data point that you found important or interesting?

For the participant(s), the first question allows them to list their thoughts so they can proceed to the following questions.

However, from a raw data point of view with respect to the post-analysis phase of the Threatcasting Method, this question allows the participant(s) to identify which of the concepts within the prompt they thought most important to the research question or threat future that will be modeled in later phases. The hierarchy of the answers in the workbook illuminates the mood or prioritization for the participants.

Additionally, how the participant(s) write their description of the data point can be instructive as well. In most cases, they may simply copy a general name or label but in some cases the participant(s) will include commentary or subtle positioning.

For example, if a data point focused on AI and labor the participant could write "A.I. and Labor" which would be a general label. But they may also write "A.I. Labor optimization" or "A.I. stealing jobs" each label can give clues to the participant's outlook and bias.

What are the implications of the data point on the threat futures or research questions?

For the process, question two is typically the most robust. Here the participant(s) explore what the data point "means" to the research question or threat future. The participant(s) can use this question to "warm up" to thinking about the future and how the data point might shape it.

This question also allows the participant(s) to discuss and capture their bias, opinions, contradictions, and expertise. At this stage in the process, this is highly encouraged. The Threatcasting Method embraces the idea that all of the participant(s) have bias, opinions, and expertise. Each of the participant(s) has been curated to be a part of the workshop because they each bring an important perspective.

In the post-analysis phase, question two can be valuable as a commentary on the data points. The data collected might expand or refine the prompt and illuminate future implications. This data can be used in addition to the threat futures that are modeled later in the process.

Are these implications positive or negative?

Question three gives further commentary to the implications. The open-ended nature of the question accepts that the implication can be positive, negative, or both. The question also raises the question: For whom is the question positive or negative?

Here we begin to bring in the personal perspective, the perspective of the people who will experience the threat. Often the prompt might be positive for one person or group of people while having a negative effect on others.

A simplistic example of this could be a development of a new weapon. For the group or country that invented and possesses the weapon the implication will be positive because they have a strategic advantage over their adversaries. However, for the adversary the effect of the prompt will be negative.

The result of the question is to get the participant(s) to explore the implication of the data points from multiple angles and perspectives.

What should we do about it?

Question four is the most open-ended. The "we" in the question is open for interpretation. The "we" could be the participant(s), a specific organization or country, or most broadly the entire human race. The participant(s) can list multiple actions for multiple parties.

This question is used to get the participant(s) thinking about specific micro actions that can be taken in response to the macro prompts. This type of thinking is essential to later phases in the process. In subsequent phases, they will not only model possible and potential threats, but they will also backcast to explore how to disrupt, mitigate, and recover from those threats.

3.1.3   DISCUSSION

The final activity in Phase 1 is to use the captured data to generate a discussion about the prompts. The data capture in the workbook is the main goal for this exercise, but an additional goal is the socialization of the participant(s) ideas to the entire workshop. To accomplish this, have the participant(s) "report out" on the data, perspectives, and opinions captured in the RSW. This "report out" brings a key component of a three-step learning and data processing process that is used in Phase 1 to a close.

Modes of Learning

When the participant(s) were presented the prompts, they were engaging in passive education.

With the shift to data capture, the participant(s) shifted to active learning. They were taking the prompts they had just heard and then applying them, thinking about the implications and envisioning possible actions.

The final step is to have the participant(s) report out or "teach" their results. This shift from active to passive to teaching allows the participant(s) to process the prompts in multiple ways.

Additionally, the reporting allows the participant(s) to hear and experience the other participant(s)' perspectives on the prompts. Following the report out, participant(s) are encouraged to discuss the similarities or differences in perspectives. This fills the room with data and perspective on the prompts.

This final stage is beneficial because it allows the participant(s) to express themselves and experience different points of view from others. It also prepares the workshop to move to the next phase, Futurecasting, where the participant(s) will use the prompts and participant(s) perspectives to craft EBM for possible and potential threats.

Interview

Dr. Erin Carr-Jordan has spent nearly two decades in higher education where her roles have included faculty, department chair, national director, and associate dean. As Head of Social Impact at Arizona State University, she drives strategy to architect, execute, and scale high-impact initiatives with a particular focus on gender equality, at-risk populations, diversity and inclusion, ethics, and interdisciplinarity. We asked Erin to talk about how people learn, take in, and process information.

When we talk about the construct of learning and acquiring new knowledge, I think there's a couple of pieces that become really important for information processing theory and attention (Atkinson and Shiffrin, 1968; Cowan et al., 2005).

How do you do things that get people's attention? And then once people are attending, it is by definition active, right? I think you're incorporating all of the different learning styles and learning modalities with the Threatcasting process, which I think are tremendously important. And I think if it were me and I were talking about acquiring new knowledge and learning a new mechanism or methodology for knowledge transfer, I would look not just to active/passive, but also the learning modalities and styles. Because from the minute you receive an invitation, there's an active engagement that you think, what is Threatcasting? That is not passive. The incorporation of auditory, visual, kinesthetic, and the weaving that you do have the experiential place (Kolb, 2014) that gets people potentially from a cognitive dissonance, I would say is likely not found a lot in the groups that you're working with, but it is a primary thing that you're trying to overcome.

If you're talking about teaching others, you have to get people to recognize that a reality exists that might not be congruent to the one that they currently have. In addition to that idea of passive to active, and it is those constructs that lie within that, that are really important. Then you're thinking, then your group brainstorming, then you're designing, developing, using your imagination together to storytell. Then you're reporting out to the rest of the room to share not only your story, but also some key things that you learned. We're trying to describe the overall benefit of all of those things taking place in a room. You're talking about really important concepts in learning theory: Why do and how do people acquire new knowledge?

I think one of the things, and I'll just speak from my experience, I find so interesting or found so interesting about Threatcasting is that it touched on all of those things and I'm going to go back to the idea of attention because it is a human brain learning construct. It is the first thing that you need to do for people to acquire new knowledge; whether you're talking about moving from active or from passive to active, the first thing that you need to do is to get people's attention. And there are certain things that when you're talking about the human brain and things that get our attention. Doing something out of the norm, which is incongruency, which is one of the constructs that flips into attention (Eccles, 1983).

I would argue that it's not necessarily passive when you start. It's the multiple modalities (Nenniger, 1992). So from a learning standpoint (Renninger et al., 2014), the way to me, it's more about motivation (Maslow, 1943). It's about learning modalities and it's about attention, and understanding that it is the weaving together to, to get to the place where you are absolutely actively engaged (Dweck and Leggett, 1988) in this. But I think it's just shepherding through the process, understanding all of the different things that we know to be true about human behavior and learning and strategically weaving them into the Threatcasting process.

3.1.4   PREPARE FOR PHASE 2

At the conclusion of Phase 1, the data in the participant(s) individual group's RSW are consolidated into a single workbook to be used as an input to Phase 2.

Phase 1 Prepares the Participants by:

- briefing them on relevant research, data and perspectives (prompts),

- focusing participant(s) attention on specific future threat area outlined in the research question,

- socializing the participant(s) with all participant(s) in the workshop and in smaller working groups (where applicable),

- giving participant(s) a platform to explore and discuss the prompts,

- capturing raw data in RSW for Phase 2 and for post analysis,

- familiarizing all participant(s) with the perspectives of the other workshop participant(s) and prompts discussion, and

- gathering all raw material and inputs needed for the Threatcasting Workbook.

3.2   POETRY

The Future

The word is the future, is the future, is future, is the future
Brother, sisters [6]

Guy

Why These Lyrics Matter

These are lyrics from the group Guy. The verses are a call for people to look around and recognize that the future may not be what we think it is. This is a lot like what groups do during the Research Synthesis portion of the Threatcasting Method. As we gather the wisdom of the crowd, we want people to challenge the future and challenge each other.

The verses finish up on a note of hope. It reminds us that we can work together to make the future better. Just as in the RSW, the future implications can be both positive and negative at the same time. Ultimately the question becomes what should we do about it?

---

[6]   Read the full lyrics at https://bit.ly/39T78II.

3.3   EXERCISES

3.1

Collaboration between people advances the level of productivity when teams are writing in the RSW. The more collaborative the team, the deeper the quality of the data. Collaborative teams develop a common language, and that aids in overcoming fear of speaking up or sharing, and personal blocks slowing down the process. Collaboration inside a team gives space for big talkers and shrinking violets, drawing people out or shutting people down as needed for the team's benefit.

Construct a checklist for building a collaborative team. Some things to consider include, but are not limited to, the following.

- What procedures should be in place to keep the Research Synthesis process moving along?

- What agreements between the team are most important?

- What policies should be in place to manage disagreements?

3.2

An essential starting part of the Research Synthesis activity is the curation of prompts. Create a list of where to find possible prompts by asking questions such as the following.

- What are the most pressing issues in the topic area?

- Who is writing about the topic area?

- Who is speaking about the topic area?

- Who is considered an expert opinion on the topic area?

- How might the topic area evolve over the next decade?

You can either use the same topic area from the exercises in Chapter 2 or pick a new topic area for this exercise.

3.3

Creating an RSW relies on your understanding of the Threatcasting Foundation and your knowledge about people. Answer the following questions.

- How might the RSW questions focus participant(s)' attention on specific future threat areas as outlined in the research question? Does this influence your word choice and visual treatment of the RSW?

- How might the design of the RSW encourage participant(s) to explore and discuss the prompts?

- How might the facilitator remind participant(s) to familiarize themselves with the perspectives of the other workshop participant(s)?

CHAPTER 4

Phase 2

"Imagination is more important than knowledge"
– Albert Einstein

4.1 FUTURECASTING

The next phase of the Threatcasting Method uses the data and perspectives captured in the RSW to envision possible and potential future threats. This futurecasting transitions from the high-level curated perspective from the prompts to a specific EBM. Participant(s) will develop this EBM by integrating elements of participatory design, experience design, and the science fiction prototyping process. This phase ends with a threat future that will then be backcasted, exploring how the future threat might be disrupted, mitigate, or recovered from.

The activities within this phase are:

- develop the EBM of future threats and

- capture all raw data.

4.1.1 EFFECT-BASED MODEL (EBM)

The central device used in the Threatcasting Method is the EBM. The EBM is a qualitative model wherein participant(s) use the prompts and the RSW to imagine possible and potential threats in the future.

Typically, a ten-year time frame is used, but shorter or longer time spans can be used. The decade in the future is used because it is "too far" out into the future. It is further out than most product cycles, political terms, or job tenures. The 10-year distance allows participant(s) to imagine a wider range of threat futures. However, 10 years is also not too far out. From a social science perspective, in the next ten years people will not change radically. Any technology that is likely to affect the participant(s) threat futures in a decade most likely exists presently, is being tested in a corporate lab, or is being incubated in a university. Ten years in the future is just far enough to free up the participant(s) to envision threat futures they may not typically consider, while at the same time using the prompts to keep them tethered to reality.

At its very basic level, the EBM is a structured database (e.g., spreadsheet) that poses to the practitioners a series of questions and requires a list of tasks to fill in in order to create the qualitative threat model. Moving from the high-level research (prompts) and participant(s)' perspectives in the RSW, practitioner(s) focus on a specific threat future. Moving from the maco-level to the micro-level, practitioner(s) explore a person in a place experiencing a threat.

This approach separates the EBM from other foresight tools such as scenario planning (Van der Heijden, 2011) and the futures wheel (Glenn, 1972).

Participatory Design

To capture this data in the EBM, the Threatcasting Method uses the practice of participatory design.

> "At the core of Participatory Design is the direct involvement of people in the co-design of tools, products, environments, businesses, and social institutions. In particular, Participatory Design has developed a diverse collection of principles and practices to encourage and support this direct involvement. Many of the design tools and techniques generated to further this process have become standard practice for the design and development of information and communications technologies and increasingly other kinds of products and services. These design tools and techniques include various kinds of design workshops in which participants collaboratively envision future practices and products; scenarios, personas and related tools that enable people to represent their own activities to others (rather than having others do this for them)" (Robertson and Simonsen, 2012).

This collaborative nature of looking into the future encourages discussion and debate, ultimately generating more detail and data in the EBM. In Part 2, Chapters 9 and 10, on Large and Small Groups, we explore the nuances and details of pulling together the groups that fill out the EBM.

Experience Design

The questions and tasks contained in the EBM draw from a range of tools, theories, and practices. Using participatory design, the participant(s) are given a series of tasks and questions to generate their future threat EBM. Threatcasting also pulls from the practices and theories of experience design. The participant(s) are not tasked with specifically identifying the future threat. They are asked to explore the effect that threat will have on a specific person in a specific location. This experience-based approach generates more detail and shifts the mindset of the participant(s).

Experience Design emerged at the beginning of the 21st century. Its origin grew out of the information technology (IT) or high-tech industry. The practice of human computer interaction (HCI) worked to define how users interacted with computers and how computers delivered information back to them (Karray et al., 2008).

This led to User Experience (UX), which further refined the desired and measurable experience that designers wanted people to have with products. Experience Design unified these ideas and applied them more broadly, beyond computer interfaces and hardware.

As we moved into the 21st century and computational systems became more ubiquitous and a part of people's everyday lives, Experience Design sought to document, explain, and design a wider range of interaction that a person might have with a product or service. This moved beyond the software or device itself to the setup process, advertising, and marketing as well as ongoing services and applications.

"The most important aspect of any design is how it is understood in the mind of the audience," Shedroff (2001) wrote in *Experience Design*, the first book written on the subject. Shedroff broadens how people are pushed to think about the design of products using an experience-based approach. As a part of that design process, he recognized that the mental model people used to understand a product or services was important for the use of that product.

"New cognitive models can often revolutionize an audience's understanding of data, information or an experience by helping them understand and recognize things they previously couldn't understand in a way that illuminates the topic or experience" (Shedroff, 2001).

Primary in these new cognitive models was the use of metaphors to build new cognitive models. In 2011, Shedroff expanded this view to say that "every sketch, model, and prototype is an elaborate fiction on the road to becoming something real" (MacWorld stage).

Threatcasting combined these approaches as the unifying principle behind the EBM. Participant(s) are not tasked with directly describing the threat. They are walked through a series of questions to describe the effect the threat will have. This helps capture a far greater set of complexity as well as second- and third-order effects of the threat that might not have been captured if the participant(s) was just focused on the single threat.

Science Fiction Prototyping

The Threatcasting Method also uses the Science Fiction Prototyping (SFP) process to broaden the participant(s) models of the future. To do this they are presented with a series of questions to help them draw on the information/data from the prompts and RSW. The answers to these questions populate the EBM with data.

"The core methodology of Science Fiction Prototyping (SFP) incorporates creative arts (for example, science fiction stories, comics, movies, art installations) as a means of introducing innovation into scientific and engineering practices, business activities and policymaking. The goal of the process is not to forecast or predict the future. SFPs focus on inventing or imaging a possible future by exploring trends from research and futurecasting.

SFPs allow organizations to investigate the human impacts that have been identified through the futurecasting process. SFPs can scrutinize the political, ethical, legal, and business impacts of

these futures. To do this the SFP process follows a simple set of rules (as illustrated in Figure 4.1)"
(Bennett and Johnson, 2016).

Most all stories (narrative plots) involve:

- a person,

- in a place, and

- experiencing a problem or threat.

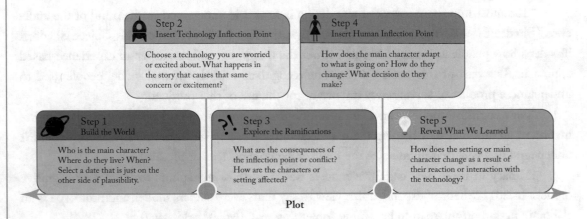

Figure 4.1: Science Fiction Prototype (SFP) five-step process (Johnson, B. D., 2011).

Following the SFP process (Figure 4.1), the questions in the Threatcasting Workbook encourage Practitioner(s) to imagine a person, in a place, experiencing a threat. From there, they are
led through a series of experience design-based questions (reviewed in detail later in the chapter)
to generate more detail and data around the threat.

After the threat is described in the workbook, the question and tasks shift to utilize a different concept. Once the threat is described, Practitioner(s) then explore the factors that brought
about or enabled the threat. This set of questions are taken from the military concept of Effects-based Operations (EBO).

Effects-Based Operations

EBO is "a process for obtaining a desired strategic outcome or "effect" on the enemy, through the
synergistic, multiplicative, and cumulative application of the full range of military and nonmilitary
capabilities at the tactical, operational, and strategic levels" (Rickerman, 2003). EBO begins with
the desired effect or outcome and works backward, employing whatever is needed to achieve the
desired outcome.

In foresight, this is similar to the futures wheel, which "... seeks to develop the consequences of today's issue on the longer-term future. We can ask how a particular new technology might influence us 20 years from now. The futures wheel does not stop at first order impacts, but rolls along to second order impacts, and beyond. It intends to explore and deduce unintended consequences" (Inayatullah, 2008).

Threatcasting harnesses the futures wheel concept for imagining and exploring but extends it beyond a single effect of a future event. This creates a more detailed EBM that ultimately explores the threat in greater depth.

In this way practitioner(s) are required to think about the threat from multiple angles, utilizing a 360° view or even a whole of society approach. This broader definition allows the practitioner(s) to model a broader range of futures while at the same time capturing a larger more detailed data set for subsequent phases of the Threatcasting Method.

4.1.2   CAPTURE THE DATA

The following section will walk you through each step in the futurecasting phase of Threatcasting (i.e., Phase 2 of the Threatcasting Method). Each step is linked directly to the Threatcasting Workbooks. The tasks and questions in the workbook make up the core requirements for the EBM that the practitioner(s) will construct. Pictures are provided of sample workbooks to help visualize the concepts presented.

Picking the Foundational Data Points for the Workbook

The intersection of the prompts and the RSW commentary around their implications give the participant(s) the raw materials to create a future that is plausible and based upon current research/science.

The first section of the Threatcasting workbook tasks the participant(s) to enter the foundational data points (curated from the prompts in Phase 1) that they will use to inform the rest of the workbook. An example is provided in Figure 4.2. The yellow highlighted areas within the workbook represent the locations that participant(s) are expected to fill out.

Data Points: Foundation of the world and the threat	
NOTE: Roll the Dice to pick a data point from each of the research areas in the Research Synthesis Workbook	
Prompt 1	
Prompt 2	
Prompt 3	
Prompt 4	
Prompt 5	

Figure 4.2: Example of Data Points in the Threatcasting Workbook.

Recall all the data points that were generated in the RSWs for each Threatcasting prompt. Now, participant(s) need to select one data point from each prompt they were presented. These will become the foundation for their vision of the future.

There are multiple ways for participant(s) to select the prompt they will use to inform their future threats. The ultimate goal is to select a range of prompts from across all of the domain areas. Additionally, for the small and large group, participant(s) should pick quickly, as to not waste too much of the workshop deciding on the prompts. The goal is to spend more time filling out the entire workbook.

In some instances, facilitators have used rolling dice, drawing cards or random number generators to push the participants to pick their prompts quickly. It also provides a degree of randomness to the prompt, as in some cases the prompts may contradict each other. This can have a positive effect for the participant(s) then need to reconcile the prompts or make a decision of what future they will use.

Details for different methods for picking the prompts will be explored in Part 2 where the application of the Threatcasting Method is explored.

Building the EBM with "A person, in a place, experiencing a threat"

After establishing a mental visualization of the environment based on the prompts and RSW, the participant(s) imagine a specific person living in that future. This adds the human dimension to the EBM. By focusing on a single person, the participant(s) isn't overwhelmed by the enormity of the threat. They are also not worried about "getting it wrong." They are simply exploring a single person, in a specific place, experiencing a threat in a decade's time.

Participant(s) are tasked to envision who the person is, their family, and the broader community with which they identify. As any good storyteller knows, no one is truly completely isolated, and so the participant(s) are asked to create the community with which their character best identifies. This community (whether physical or digital) provides them the support during their upcoming trial and will also be whom they tend to turn to.

The participant(s) then explore where the person lives through a series of questions. An example of these questions is shown in Figure 4.3.

Depending upon the audience and the research goal, participant(s) can be prompted to fill in minute details about their person's surroundings and daily life as well as their hopes and dreams.

The next step is for the participant(s) to describe the Event. The physical or digital instantiation of the problem caused by the threat is the "event." To better model and understand the event, the participant(s) are asked a series of questions that are designed to push the them to add as much detail as possible to explore and explain their futures.

PART ONE: Who is your Person?	
NOTE: Remember to give as much detail as possible. The power is in the details.	
Who is your person, what is their name, and what do they look like?	
Who are their friends and what is their broader community?	
Where do they live? What is their occupation?	
What is the threat to their way of life?	
Briefly describe how your person experiences the threat (The Event) and possible 2nd/3rd order effects.	
Who else in the person's life is involved?	
What specifically does the Adversary or Threat Actor want to achieve? What is the Adversary or Threat Actor hoping for? What is the Adversary or Threat Actor frightened of?	
What vulnerabilities does this expose?	

Figure 4.3: Example "person" questions in the Threatcasting Workbook.

Going beyond the "5Ws" of traditional information gathering (who, what, when, where, why), Threatcasting questions are specifically designed to create a more well-rounded narrative describing the complexity and uncertainty of the threat and operating environment.

The description of the event can use a more traditional scenario-planning process (Ringland and Schwartz, 1998).

Example Questions

- Who is your person and what is their broader community?

- What is their occupation? What does a day in their life look like?

- Where do they live?

- What is the Event?

- Imagine your person experiencing the Event.

- What is it? Who else in the person's life is involved? What does the Adversary or Threat Actor want to achieve? What is the Adversary or Threat Actor hoping for? What is the Adversary or Threat Actor frightened of?

Interview: "How to Imagine the Future"

August Cole is a writer, analyst, and consultant specializing in national security issues. He is a nonresident senior fellow at the Brent Scowcroft Center on International Security at the Atlantic Council. He is the director of the Art of Future War project, which explores narrative fiction and visual media for insight into the future of conflict. His novels Ghost Fleet *(2015) and* Burn In *(2020), co-authored with P. W. Singer, are considered essential reading across security and defense professionals in the military, government, and private sector.*

We asked August about his process for imagining the future and some simple rules to follow when creating a threat future.

Remember: The Future is About People

One of the foundational aspects to creating stories about the future is to remember they are environments that will be inhabited by people. As much as it is tempting to have technology drive everything from a scenario to plot or character development, it's important to fundamentally remind yourself and also your reader that what they are learning about ultimately is a very human experience.

The research you will use to drive the story idea will encompass an enormous amount of learning about future technical systems and demographic research. But all of that research exists within a construct to create an environment that feels believable.

To See the Future, Remember the Present

One of the rules I used in writing *Ghost Fleet*, *Burn In*, and the stories that I've written about the future of conflict is that technologies within the future story need to be drawn from systems, capabilities, or products that exist today or are in development. This is useful because you're asking the reader to stretch the bounds of their imagination. You have to push the bounds of the imagination to arrive at something innovative. If you've done your research and you've based your future world on trends, technologies, ideas that are contemporary or on the horizon, then can push the envelope.

In *Ghost Fleet* we used a fairly outlandish scenario. We had the ruling regime of China in the late 2020s take the audacious step of seizing Hawaii. We wanted to show in an instant that the world order was no longer the same. By taking that step the future will be forever changed for the rest of the 21st century. To make the scenario seem plausible and real we used technologies like swarming drones, AI, and battlefield exoskeletons on soldiers. These were all tech that was real or in development by the Chinese or Allied arsenals. The research we put into the book would be just the same as if we had written a nonfiction book.

Create Useful Fiction

You have to remember that these futures are a kind of useful fiction or science fiction prototype. I've used a somewhat tongue-in-cheek term for them: FICINT—fictional Intelligence. It's similar to existing terms like HUMINT, which is gathering intelligence using humans or people. Or SIGINT, which is using signals to gather intelligence. FICINT, uses fiction to gather intelligence about possible futures. There is value in collecting, understanding, and acting on the kinds of future narratives we write.

Consider what your reader or the user of the scenario is missing in their assessment or assumptions about the future. Always try to surprise them, don't reiterate or reinforce conventional wisdom. When you are thinking about a concept, ask yourself is there a point of view or a character who is going to be overlooked by the average reader. In fact, if you don't do your research and understand where the outlier ideas are that are possible, then you can examine the underlying issues to a threat or technologies that could give one group a strategic advantage over another.

What's the Ask?

When I create these futures, I'm always thinking about what's the ask? What am I trying to accomplish? When a person reads the scenario, what do I want them to do differently? Do I want them to be more aware of a threat or think with more criticality about their role in a given situation? Do I want them to understand they have the possibility to change an outcome they might think is predetermined? The notion of "what is the ask?" is important so that you can arrive in a few words at a meaning and impact that is much bigger than the vessel the story or scenario represents.

Experience Questions

The next section of the Threatcasting workbook draws directly from the practice of experience design. To gather detailed information for the EBM, participant(s) are asked a series of questions to expand their description of the person's experience with the threat and the event.

These questions should be curated to gather the correct data needed to answer the research question. Also, it is important to understand the participant(s) and use the questions to expand their thinking or challenge them.

The nature of the experience questions is different from those asked previously about the person. There, the questions are more open ended. Participant(s) are tasked to focus on thinking of a character and a place, describing them and the event. The experience questions are meant to probe the participant(s) to describe what the event might look or feel like. These are experience questions. To describe what it would be like to experience the event from the person's perspective.

This type of question is important because the participant(s) are describing and modeling the effect the threat will have on the person and their community. These effects will be necessary for the next step in the Futurecasting process—the enabling questions.

PART TWO: Experience Questions (from the perspective of "the person" experiencing the threat)	
Questions (pick two)	
What will this make your person do that they normally would not?	
What is different and/or the same as previous events or instantiations of the threat?	
When the person first encounters the threat, what will they see? What will the scene feel like? What will they not see or understand until later?	
How will information be delivered to the person? Where and how will the person connect and communicate with others?	
What will the person have to do to access people, services, technology and information they need?	
What is the worse case scenario?	
What are the broader implications of a threat like this? What might a ripple effect look like?	
Question One	PASTE the question here; answer in yellow box
Question Two	PASTE the question here; answer in yellow box

Figure 4.4: Example of Experience Questions in the Threatcasting Workbook.

Depending upon the size of the threatcasting session and the amount of data needed, the participants(s) can be tasked to answer all of the experience questions or a subset if time is a factor. Figure 4.4 provides an example of what these questions might look like in the Threatcasting workbook for participant(s).

Example Experience Questions

1. What is different and/or the same as previous events or instantiations of the threat?

2. When the person first encounters the threat, what will they see? What will the scene feel like? What will they not see or understand until later?

3. How will information be delivered to the person? Where and how will the person connect and communicate with others?

4. What new capabilities enable the person and their broader community to recover from the threat?

5. What are the broader implications of a threat like this? What might a ripple effect look like?

6. What is the worst-case scenario?

Interview: "What is Experience Design?"

Nathan Shedroff speaks and teaches internationally and has written extensively on design and business issues. His 2001 book, Experience Design, *was the first book published on the methodology of experience design. He's a serial entrepreneur, works in several media, and consults strategically to help companies build better, more meaningful experiences for their customers.*

We asked Nathan to provide more detail about experience design and the six dimensions that make up the approach.

The History of Experience Design

I come from a car design and architecture background. Everyone that comes into the experience design field comes from a slightly different place, which is why it's been so rich for so many years. This field comes out of a rich tradition, starting with what was called human factors back in the 1940s to the 1970s. Human factors involved engineering for people, growing out of the military and cockpit design. Later for our purposes it was adopted in software development and typically called usability or user testing in the 1990s.

These products were built by computer scientists and engineers without a lot of understanding or care given to the people who would be using them. This kicked off a plethora of investigation and application of these ideas that came from graphic design, industrial design, architecture, marketing, and even branding. What happened was that the field exploded, taking and looking to solve a wide range of issues and impacts. It attempted to integrate or address in the design of pretty much everything. I turned my attention to the concept of experience to encapsulate the various areas where work was being done.

Ultimately, experience design is an approach, as opposed to a discipline. You can be a car designer and you can be an experience designer. You can be a graphic designer or an events designer, or a fashion designer or an architect, and be an experience designer at the same time. To be an experience designer you need to open your design process to all of the elements of experience, not just the ones traditionally focused upon in a specific field.

The Six Dimensions of Experience Design

1. Breadth

Ask yourself: What is the full breadth of the experience? Try not to focus on a single device or application. You look across all the touch points and people that are involved: users, customers, and constituents. This could be a physical product or service. In fact, every product pretty much has services attached to them. They can also be spaces, places, whether that's a gallery, a home, a workplace, and a theme park. They can be events or music concerts. Any place where people experience a broader interaction.

2. Duration

Everything happens in time. Everything has a duration. Even if you're designing a product, someone is using it in time. By applying the experience design approach, a designer or person would examine the whole timeline. How does someone come to this moment? What is the difference between the state of mind they were in before and the state of mind once they start into the experience? What happens during the time they're using the product? How does it end? How do you gracefully get them off to the next thing without it being abrupt? How do you get them to come back and repeat?

Looking at duration is an important part of experience.

3. Interaction

Another dimension is interaction. Meaning how passive or how active, or how interactive should it be. And not everything should be wholly interactive.

4. Intensity

What is the intensity of the experience? This can be measured on a scale. On one end there is a reflex, wherein the experience doesn't even trigger your consciousness before it's over. In the middle of the scale is habit. Meaning, it is an action or interaction that a person used to be engaged in, but now they don't think about it. Finally at the other end of the intensity spectrum is engagement. Wherein people consciously think about the experience they are having.

An experience design can move across this entire spectrum.

5. Design

This is the traditional design area that most people know. Designers make sensorial decisions about a product or service. What does it look like? What colors? What is the typeface? What is the layout? What's the form? What does it feel like? What are the materials, sounds, music, voice, etc. For traditional design professionals (like food, product, or graphic design), this comes from a set of decisions people and designers make along the way that result in the final solution. Designers are taught to reimagine the world in the image they thought they'd liked.

The experience design approach flips this concept on its head. It challenges that a designer needs to understand what triggers the audience, users, constituents, or customers. Once that is understood, then curate these decisions in such a way as to trigger the experiences we want them to have. We call these dimension triggers because each one of those decisions triggers a reaction in the people that we're trying to serve.

For example, imagine you are the CEO of a company and you want to make your wood products more green. Perhaps your first idea is to make the product out of bamboo. But the experience design approach tells us that bamboo doesn't trigger the same things in all people. In a certain customer set, bamboo might trigger green, regenerative, and sustainable connotations. But in another customer set, bamboo might trigger low quality, third world, cheaply made, even hippie like. It's up to us as designers to research and understand how customers are triggered by these different choices. This will allow them to make better, more appropriate choices.

6. Value

When we design things, as we create an experience, we're creating value. There are five kinds of value.

Note: for the purpose of this list we will talk about the values as they might pertain to a designed physical object. However, these same values can be applied to any designed object whether it be physical (e.g., a manufactured device), digital (e.g., an app or software program), imagined (e.g., a fictional story or movie), or structural (e.g., the design of an organization).

- Functional value

    ○ Does the object serve a purpose? Does it solve a problem or help with a task? Can a person use the object easily? When the person first approached the object did they know how to use it?

- Financial value

    ○ Does the object make money or enable others to make money? Are the other values such that a person would pay for the object?

- Emotional value

    ○ Does the object make a person feel a certain way? Do they become attached to the object? Does it make them laugh? Do they think it's cute? Does it spark nostalgia because the object relates to other objects in the person's past?

- Identity value

    ○ Does the object give the person a sense of belonging to a broader group? Does it make them feel like they are a part of something larger than themselves?

- Meaningful value

    ○ Does the object stand for something else? Does the object hold a broader significance beyond itself? Is it purposeful?

All of these values are desirable attributes that we could decide to make important. This affects the value that is provided to the experience. This connects experience design to business strategy and goals.

It underpins the whole purpose of the experience in the first place. Why does this experience exist? What value does it provide to people? Everything else follows along that formula as a tool to make that value happen.

If you are looking across all these six dimensions, all these pieces, you're an experience designer. If you're only looking at a small subset and not considering these other things, you still may be a fashion, interior designer, or car designer, but you're not an experience designer.

Enabling Questions

Then, as in all good science fiction stories, the perspective changes and the "event" is seen from the adversary's perspective. Who is this adversary and how did they become who they are today? What are the motivations that caused them to instantiate the threat? Truly, they had a reason (which may or may not have started out as morally, ethically, and legally sound) for their actions. The participant(s) are forced to think about the event from their adversaries' perspective to articulate what makes them tick—and often, to challenge their initial reaction that they/them are evil. Some groups come to the realization that their "adversaries" might start, ten years previously, as seeming reasonable.

The final step in the futurecasting phase of the Threatcasting Method is to understand what forces might enable the threat to happen. The nature of the tasks and questions shift from experience-based to effects-based. Drawing from EBM, participant(s) are tasked to explore a wide range of factors that could enable the threat or threat actor.

Again, the participant(s) are tasked to change their perspective. They move away from thinking about their main person and now think more objectively about the threat. This is where the participant(s)' own experience, expertise, and even biases become an asset to the process. Some of this information (strategic level or big picture) would have been captured in the RSW but the tactical areas that need to happen are better defined during this part of the Futurecasting phase.

For some it is helpful to think about this question from the threat actor's point of view or how participant(s) might enable the threat actor to bring about the threat.

PART THREE: Enabling Questions - Adversary or Threat Actor (from the perspective of "the party" bringing about the threat)	
Questions (pick two) from the drop down selections	
Barriers and Roadblocks: What are the existing barriers that need to be overcome to bring about the threat? How do these barriers and roadblocks differ geographically?	
Business Models: What new business models and practices will be in place to enable the threat? How is it funded?	
Research Pipeline: What technology is available today that can be used to develop the threat? What future technology will be developed?	
Ecosystem Support: What support is needed? What industry/government/military/criminal elements must the Adversary or Threat Actor team up with?	
Centers of Gravity: What are the sources of power that provides moral or physical strength, freedom of action or will to action?	
New Practices: What new approaches will be used to bring about the threat and how will the Threat Actor enlist their broader community?	
Training and Outreach: What training is needed to enable the threat? Where can the Threat Actor get this training/education from?	
Question One	Paste question here; answer in the yellow box below
Question Two	Paste question here; answer in the yellow box below

Figure 4.5: Example of Enabling Questions in the Threatcasting Workbook.

Depending upon the size of the Threatcasting session and the amount of data needed, the participants(s) can be tasked to answer all of the experience questions or a subset if time is a factor. Figure 4.5 provides an example of how these questions might be presented to the participant(s) in the Threatcasting workbook.

The enabling questions can be broken down into high-level categories to explore all aspects of the EBO approach (Kyle, 2008).

For example, participant(s) are tasked to explore potential roadblocks or barriers and think about new business models and practices that would enable the event. Namely, the things that could occur in the future that, ultimately, would be advantageous to the adversary. The important thing is that they look at both sides of the threat; for instance, what are the things that will need to be invented or improved over the next 10 years to facilitate the adversary's actions and access to systems? Then they also have to contemplate what barriers could be put into the adversary's trajectory that could slow them down. Participant(s) also imagine what socio-technical systems and discrete technologies would help facilitate the threat and what support systems are required for it to thrive.

Example Enabling Questions

- Barriers and Roadblocks

  ○ What are the existing barriers that need to be overcome to bring about the threat?

  ○ How do they differ geographically?

- New Practices

  ○ What new approaches will be used to bring about the threat and how will the Threat Actor enlist their broader community?

- Business Models

- ○ What new business models and practices will be in place to enable the threat?

- Research Pipeline

  - ○ What technology is available today that can be used to develop the threat?

  - ○ What future technology will be developed?

- Ecosystem Support

  - ○ What support is needed?

  - ○ What industry/government/military/local partners must the Adversary or Threat Actor team up with?

- Training and Outreach

  - ○ What training is needed to enable the threat?

  - ○ How will the Threat Actor educate others about the possible effects of the threat?

In some cases, the participant(s) think about the training necessary to enable this threat. For example, a military perspective might visualize this as exploring the entire DOTMLPF-P (doctrine, organization, training, materiel, leadership and education, personnel, facilities, and policy) spectrum (Joint Chiefs of Staff, 2021). Ultimately, the participant(s) develop the cornerstones of the adversary's strategy and/or roadmap to threat success.

DOTMLPF-P (Defense Acquisition University, n.d.**)**

- Doctrine: the way we fight (e.g., combined air-ground campaigns, multi-domain operations)

- Organization: how we organize to fight (e.g., brigades, divisions)

- Training: how we prepare to fight tactically (basic training to advanced individual training, unit training, joint exercises, etc.).

- Materiel: all the "stuff" necessary to equip our forces that DOES NOT require a new development effort (weapons, spares, test sets, etc., that are "off the shelf," both commercially and within the government)

- Leadership and education: how we prepare our leaders to lead the fight (squad leader to 4-star general/admiral—professional development)

- Personnel: availability of qualified people for peacetime, wartime, and various contingency operations

- Facilities: real property, installations, and industrial facilities (e.g., government-owned ammunition production facilities)

- Policy: DoD, interagency, or international policy that impacts the other seven non-material elements.

Interview

Dr. Andy Hall is an Associate Professor of IT, Data Science, and Cybersecurity at Marymount University. A retired Army Colonel, Dr. Hall has served as an infantry officer, an operations research and systems analyst, a cyber officer, and an academy professor. He has served in infantry and artillery units as well as multiple assignments in the Pentagon, serving on both the Army Staff and the Joint Staff.

Using his knowledge of military planning, we asked Andy to explore Effects-Based Operations in general and how they apply to the Threatcasting Method.

Effects-Based Operations—A Different way of Thinking

As an analyst using the Threatcasting Method it's important to realize that you are going to be using a lot of imagination.

When you're thinking about creating an effect, you have to spend a little bit of time in an imagination exercise to come up with multiple ways you could see something unfolding. When we were working in effects-based operations within the military and doing campaign planning often we would have a specific military goal or mission. But we were trained to then question: What else could we do? What are the other things could we add? What are the ways that we could influence a local population to make our job easier? You start to come at the problem from multiple angles, imagining what could be done to reach your ultimate effect.

Effect-Based Operations at Work in the Threatcasting Method Enabling Questions

A lot of effects-based operations are trying to search for the center of gravity for a situation. What is the one effect you could achieve that would help you achieve all of your goals? The Threatcasting Method makes this very clear really quickly. By describing the threat in detail, the analyst(s) knows that the goal is to disrupt, mitigate, or recover from the threat. But also, the analyst(s) can think about the EVENT, the physical or digital instantiation of the threat as the effect that the adversary or bad actor is looking to achieve.

Then they can think about all the different steps that were taken to bring that threat about. It's a funny way of thinking for a military person. You are asking me to imagine how to enable the enemy. But the enabling questions give you a specific and tangible way to thwart the threat from happening.

One way to break this down is to use elements of national power: the diplomatic element, the information element, the military element, or the economic element. But as an analyst you get to imagine a wider variety of elements that enable the effect. Everything from technologies, relationships, communication, and popular culture.

The important part is to think through all of the different aspects that could have enabled the effect, enabled the threat and the event to happen. The more specific you are the better.

Words Matter—Make Sure You're Speaking the Same Language

When we were in Baghdad, we brought in advertising executives to talk to our commanders. A lot of what we were trying to achieve was to get a lasting effect with the local population using information and cyber operations. We realized the effect we wanted was to change people's minds and opinions. The effect we wanted could be achieved by thinking like people who specialized in advertising and brand marketing.

When we brought in these executives, we quickly learned they did not speak the same language as the operational commander(s). When they tried to explain their pitch, no one in the room had a business background and they couldn't understand. They didn't realize that they actually needed a translator, somebody that could explain the idea in the language for the military and for the advertising people.

As you begin to populate your EBM, it's important to understand the language of the people you are trying to affect and also the people who will be using the information in the Application Areas. It's also important to remember as you collaborate and work with other analyst(s) and participant(s).

Systems Thinking and the EBM

EBMs can become complex, with a number of experience design and effects-based operations elements. Systems Thinking can be helpful too for the analyst's well-rounded understanding of the EBM. Systems Thinking is a way of experiencing all the things around us, and turning them into ideas and narratives that can be broken into pieces, analyzed, and deeply looked at for how they intersect and influence all other things.

In one lecture, the nobel laureate and physicist Richard Feyman (2006) described an approach to systems thinking:

> "A poet once said, 'The whole universe is in a glass of wine.' We will probably never know in what sense he said that, for poets do not write to be understood. But it is true that if we look at a glass of wine closely enough, we see the entire universe.

There are the things of physics: the twisting liquid, which evaporates depending on the wind and weather, the reflections in the glass, and our imagination adds the atoms. The glass is a distillation of the earth's rocks, and in its composition we see the secrets of the universe's age, and the evolution of the stars.

What strange arrays of chemicals are in the wine? How did they come to be? There are the ferments, the enzymes, the substrates, and the products. There in wine is found the great generalization: all life is fermentation. Nobody can discover the chemistry of wine without discovering the cause of much disease. How vivid is the claret, pressing its existence into the consciousness that watches it!

If in our small minds, for some convenience, we divide this glass of wine, this universe, into parts—physics, biology, geology, astronomy, psychology, and so on—remember that nature does not know it! So let us put it all back together, not forgetting ultimately what it is for. Let us give one more final pleasure: drink it and forget it all!

"What makes this famous quote an example of systems thinking is the way he transforms contextual patterns: he transgresses parts and wholes, takes new perspectives, forms new relationships, and makes new distinctions…" (Cabrera, Colosi, and Lobdell, 2008).

Systems thinking can be applied to any complex system or concept like the EBM. This way of thinking allows the analyst(s) to understand the complexity by breaking down the system into parts.

Systems Thinking Example: "Know thy Self"

How might Systems Thinking be applied to a complex system so that the analyst(s) can break that system down and understand all of the parts that make up the whole? A good example of a complex system is humans. What could be more complicated than people?

Systems Thinking has been used in the medical field to improve patient outcomes in a complex healthcare delivery system. To do this, researchers developed a "helix" or complex systems map of the human self.

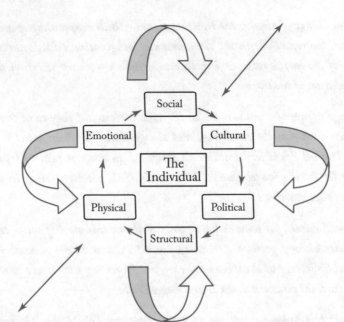

Figure 4.6: Continuum of Complexity Example showcasing System Thinking. (Stalter et al., 2017. Courtesy of Wiley.)

Figure 4.6 "… provides a view where whole individuals exist within the center of the helix. In essence, systems thinking is what occurs when the individual's social, cultural, physical, emotional, and political attributes change the system but at the same time are changed by the collective nature of the system. The lines are an intersection of time. Time infers individuals have levels of experience and familiarity with the system. So, as one comes to know the system, one comes to know the self" (Stalter et al., 2017).

This approach gives the analyst(s) a way to take the complex system, think about the composite parts that make up that system and document them. This example is helpful as well because it allows the analyst(s) to explore how the different parts of the system might interact with each other over time.

4.2   POETRY

The Future—never Spoke

The Future—never spoke—
Nor will He—like the Dumb—[7]

Emily Dickinson

---

[7] Read the full text at http://archive.emilydickinson.org/working/h336.htm or https://bit.ly/3mccnJo.

Why This Poem Matters

In Dickenson's poem, that Future refuses to reveal itself to people. She expresses that the future can't be known and that the future only reveals itself "in the act," meaning that the future is revealed by experiencing it. Just as in Threatcasting, we are not predicting the future; we are not trying to reveal the future. We are exploring a range of futures—not a single future but many. And the way in which we explore these futures is by experiencing them "when the News be ripe—Presents it—in the Act—"

In Dickenson's poem, that Future's job (given a male persona), his "office" is to execute fate's telegram. It is interesting to think about the futurecasting in the EBMs as telegrams from the future, or rather delivering these future telegrams to our fellow participants.

4.3 EXERCISES

4.1

"The only unique contribution that we will ever make in the world will be born out of our creativity" —Brene Brown.

To build out your person in a place experiencing a problem takes imagination and creativity. This exercise is a skill builder for flexing your imagination. Let your imagination play and be creative when taking the following action.

Use elements that speak to human experiences, such as describing through the five senses. You can place your "person" in the place much easier if your reader can smell the farmland surrounding the person, imagine the dusty soil's feel, and recall the farm animals' sound.

Create your person in this story by completing the following.

1. Write a descriptive and creative name for your person.

2. Write five words to describe your person; use your senses.

3. Write the neighborhood, city, state/territory/province, country where your person lives.

4. Write a few sentences describing what's happening to your person, what action is taking place? What are they experiencing?

5. Write three to five lines of dialogue, placing your person in a conversation.

6. Review all of the above and make each one even more descriptive.

4.2

The following exercise describes and breaks down 3 of the 6 Dimensions of Experience Design.[8] Each dimension has elements that distinguishes and differentiates experiences. The questions and descriptions in this exercise will allow you to examine the experience a person might have from multiple points of view.

First, pick a product (e.g., car, clothing, furniture) or software application (app, program) that you have interacted with in the past. Recall how you encountered, explored, and interacted with it. Use the following example dimensions to get a better understanding of the experience design that went into that product or software application.

Value

The most important dimension is the kinds of value the experience provides to people.

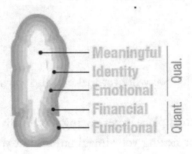

Figure 4.7: Values in Experience Design Thinking. (Shedroff, N., 2001. Courtesy of New Riders.)

NOTE: Those values closer to the center are both more stable and more "valuable." They are shown as lighter colors in Figure 4.7.

1. What are people's decision-drivers for each segment of value (e.g., meaningful, identity, emotional, financial, or functional)?

2. How do people find value in the product?

3. What parts do they value and why?

4. Are there certain values that are stronger or a higher priority than others?

Triggers

When a product is designed there are specific choices that can trigger a reaction in people. For this product/app, what reactions are triggered in the people you're designing your selected product/app for?

---

[8] Exercise provided by Nathan Shedroff (https://nathan.com/tool-experience-model/). It is meant to explore some of the "6 Dimensions of Experience Design."

- Sight

  - What does the product look like?

  - Is it designed to look a certain way or elicit a specific feeling?

- Sound

  - If the product makes a sound, what is the quality of that sound?

  - Is it assertive, pleasing, urgent, relaxed?

- Touch

  - If the product is physical, what does a person feel when they touch it?

  - What material is it made out of?

  - What is the quality of the materials?

  - Does it feel luxurious or frugal?

- Smell

  - If the product has a smell, how is that smell designed?

  - Is it floral or woody?

  - How would you describe the smell and the desired effect it could have on a person?

- Taste

  - If the product has a taste, what is the quality of that taste? Is it sweet, spicy, complex, or simple?

- Symbols

  - Are there symbols used on the product for people to understand how to use it?

  - Imagine all the different symbols that are used in a car or kitchen appliance. Each of these visuals affects the experience.

  - What do the symbols say about the product as a whole?

  - How do the symbols add to the all over experience of the product?

- Name

- ◦ What is the name of the product?

- ◦ How will people refer to it?

- ◦ What is the brand?

- ◦ How will the name make people feel?

- ◦ Is it playful (e.g., Funyuns), serious (e.g., Iron Net Security), or perhaps doesn't "mean" anything at all (e.g., Google)?

- Price

  - ◦ What will the product cost?

  - ◦ What does that cost say about the product?

  - ◦ Will it be inexpensive so everyone can have it?

  - ◦ Will it cost a lot of money so that it is a luxury good?

  - ◦ Will it be given away for free?

Duration

All products are used "in time" by people. The amount of time people spend experiencing a product can vary widely. How they interact and experience the product can be mapped across a continuum through time in which people enter, inhabit, and then exit an experience in four distinct phases (example is shown in Figure 4.8). For your product, answer the following questions about how people will use the product over a period of time.

1. Initiation

   a. How will they first encounter the product?

   b. What is their first use of it?

2. Immersion

   a. As people continue to use the product, how will they understand it?

   b. Will they get better at using it over time?

   c. Is it so simple that they will be instantly immersed?

3. Conclusion

   a. How will people finish using the product?

   b. What will "shutting it down" or "putting it down" feel like?

4. Continuation

   a. When a person is done with the product what will they think about it?

   b. How will they return to the product?

   c. How and why will they use it again?

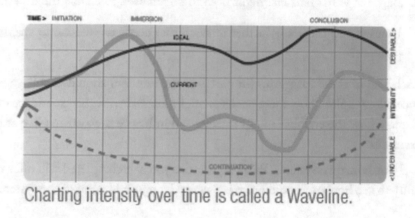

Charting intensity over time is called a Waveline.

Figure 4.8: Visualization of how time spent with a product changes in Experience Design Thinking. (Shedroff, N., 2001. Courtesy of New Riders.)

4.3 Applying Systems Thinking Example: "Know thy Self"

Use the "Know Thy Self" example of systems thinking (Chapter 4 Enabling Questions) to break down another complex system. Visualize the example helix (Figure 4.6) as a template in order to understand the interconnectedness of complex systems. Start with an experience, then "zoom-out" from there and explore the components that enable that experience to happen. Your experience could be as simple as ordering products on Amazon or as complex as learning how to drive a newly purchased car. It could be going to a doctor's visit that starts with making the appointment via the phone or the internet, parking and entering the office, talking with the staff and the doctor, going to get blood drawn for lab work, and finally the follow-up conversation with the doctor once the results have come back.

   Directions:

1. Write your experience in the center of the page.

2. Write the following enabling categories in spaces in the circles around your central experience.

    a. Economic: What economic or business models will enable the experience?

    b. Technical: Technical enablers that will help bring the experience about.

    c. Ecosystem: Who might be the broader range of people or groups that will help enable the experience?

    d. Barriers: What barriers might need to be removed to enable the experience?

    e. Research: IS there any further research that needs to be done to enable the experience?

3. Each one of these categories (in part 2 of the exercise) should enable your experience. Now explore how these enablers might interact with the experience over time. Explain how the enablers might interact and influence each other. Give at least two examples.

Think about how this Systems Thinking exercise you just completed will offer the analyst(s) the opportunity to see how each experience is enabled by multiple parts of the system.

CHAPTER 5

Phase 3

"Someone's sitting in the shade today because someone planted a tree a long time ago."
—Warren Buffett

5.1 BACKCASTING: TIME-PHASED, ALTERNATIVE-ACTION DEFINITION

The third phase of the Threatcasting Method is an expanded backcasting. Using the threat future (EBM) that was created in Phase 2, participant(s) next explore how to disrupt, mitigate, and recover from the threat. They will also identify the actions that can be taken and the indicators (flags) that will indicate the threat is beginning to manifest itself.

This kind of "backcasting" varies from traditional backcasting in its scope and complexity. Participant(s) will use a time-phased, alternative-action definition (TAD) process. TAD allows participant(s) to explore multiple time-based futures, with multiple and parallel possible actions enacted by multiple parties in different locations to disrupt, mitigate, and recover from the future threats they have identified. These multiple and parallel actions, parties, and locations can also apply to the indicators (flags) that a threat is beginning to manifest in various places, with different parties taking or showing different actions.

The output of this phase is the final portion of the raw data that the analyst(s) will use in the post analysis, synthesis, and findings phase (Phase 4).

The activities within this phase are:

- review the EBM and the "Event;"

- develop Gates and Flags;

- create action milestones; and

- wrap up.

5.1.1 INTRODUCTION TO TAD BACKCASTING

Backcasting is a technique developed in the 1970s by Lovin and coined by Robinson (1982). Simply, it is a technique used to describe a desirable future and then work backward to develop the steps for how to get there. This was in direct contrast, at the time, to strategic planning techniques that focused only on likely futures and how to achieve them from the present.

Robinson (1990) explains that "the major distinguishing characteristic of backcasting analyses is a concern, not with what futures are likely to happen, but with how desirable futures can be attained. It is thus explicitly normative, involving working backward from a particular desired future end-point to the present in order to determine the physical feasibility of that future and what policy measures would be required to reach that point" (pp. 822–823).

The Threatcasting Method expands traditional backcasting from looking at a single path to a specific threat future to a vision that explores multiple paths and possibilities. Participant(s) use a time-phased, alternative-action definition (TAD) process. TAD allows participant(s) to explore multiple time-based futures and actions from multiple parties (academia, government, military, industry, nonprofits, etc.) to disrupt, mitigate, and recover from the future threats they have identified.

A kind of pluralistic backcasting that explores multiple futures has been used in other futures studies, research, and techniques. Anu Tuominen et al. (2014) describe a form of pluralistic backcasting where "multiple visions of the future are developed in a participatory, interdisciplinary process using the Delphi method. Further, the pathways to the alternative visions are constructed."

They go on to state that this pluralistic approach has also been used in spatial planning, sustainable consumption, and hydrogen futures (Höjer, Gullberg, and Pettersson, 2011; van de Kerkhof, Cuppen, and Hisschemöller, 2009; Green and Vergragt, 2002).

The "variety of time-frames" in the definition refers to the fact that participant(s) will be describing the multitude of actions that will need to occur both within the time frame of 0–4 years and then in 4–8 years that will facilitate (or deny) the proposed vision of the future. This will help put the actions in a rough timeline and allow for exploration of the second- and third-order effects of the earlier actions onto the later actions.

The Threatcasting Method utilizes the TAD Backcasting methodology by asking participants to work backward in time from their one established future (EBM) to identify what could be done to disrupt, mitigate, and/or recover from their defined threat by way of multiple actions and multiple parties. TAD allows participant(s) to explore multiple time-based futures, with multiple and parallel possible actions enacted by multiple parties in different locations to disrupt, mitigate, and recover from the future threats they have identified. The possible multiple and parallel actions, parties, and locations can also apply to the indicators (flags) that a threat is beginning to manifest in various places, with different parties taking or showing different actions.

Traditional Backcasting

Traditional backcasting can be visualized by a single line of actions and indicators running backward from the threat future to the present (Figure 5.1). Traditional backcasting was originally explored by Lovin and Robinson in general. There, the analyst(s) envisions a possible and potential future. In the traditional backcasting, the analyst(s) consider what needs to happen to encourage the envisioned future to happen. They ask the following questions.

- What could bring this future about?

- What changes in policy might need to happen?

- What new technologies need to be developed?

- What investments need to be made?

Figure 5.1: Visual of Traditional Backcasting.

The Threatcasting Method differs from traditional backcasting in a key way. Threatcasting is preventative. It asks participant(s) to explore not only what might enable the possible and potential future to happen but also what needs to happen to disrupt, mitigate and prevent it.

In Figure 5.1, the future (represented as an open circle) is defined and a single path leads back to the present to enable that future. Gates (actions) and flags (indicators) could occur along a single line. Participant(s) consider the Enabling Questions that helped to bring about the threat.

A Future Threat Prevented

The Threatcasting Method differs from traditional backcasting in its preventative nature and the goal to identify possible and potential threats in the future as well as the specific gates and flags to disrupt that threat future. Figure 5.2 illustrates how multiple flags and gates can happen along the timeline. The first two flags (A and B) indicate that the future threat is beginning to manifest where the last gate (C) stops the threat future (D) from happening.

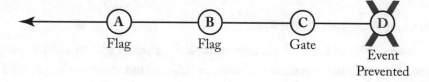

Figure 5.2: Flags (A, B) and gate (C) that prevents a threat future (D).

Branching Future

A branching future can occur when a threat happens in multiple, different locations. This will branch the backcast so that the timeline begins to have two parallel paths, occurring in two different locations. Note that location need only refer to the concept of a physical terrain location. In Figure 5.3, one of the future timelines is disrupted where the other is not.

This branching does not have to be just confined to place or location. It can also apply to multiple actions or people (actors) who take those actions and create parallel timelines for the threat future.

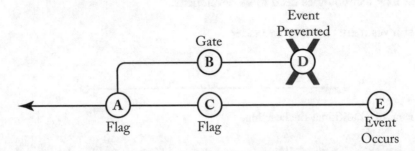

Figure 5.3: Visual of a Branching Future. A = a flag that indicates the threat future has branched (e.g., occurring in two locations differently). C = a flag that indicates in this location the threat future (E) is happening. B = gate that stops threat future D from happening.

An example of a Branching Future is a policy of law (flag) passed in one country that is fundamentally different from another. The European Union might pass a law regulating or preventing the development and deployment of AI for use in calculating sentencing recommendations for people convicted of certain crimes. At the same time, the U.S. chooses not to enact a policy or pass an equivalent law.

When backcasting the future use of AI and algorithmically recommended sentencing, the futures will branch because the use of the technology in the different locations will be different. In the EU the use of the technology will have been prevented while in the U.S. the backcasting will continue.

Parallel Paths and Pluralistic Threat Futures

Another illustration, Figure 5.4, explores how multiple flags across various actions, people, and locations can create two parallel timelines that occur at different rates and may require different gates or might present via different flags.

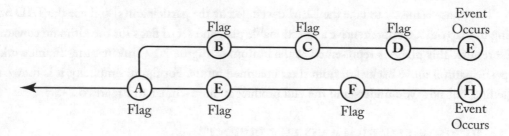

Figure 5.4: Flag A indicated a branching future. Flags B, C, D indicate threat E is about to happen on the top line. Different flags F, G in a different location indicate that a different threat future H is emerging.

TAD Threat Futures

Ultimately, the Threatcasting Method's TAD approach to backcasting creates a complex and varied map of possible gates and flags (Figure 5.5). Here multiple gates and flags are mapped across multiple possible actions, people, and locations.

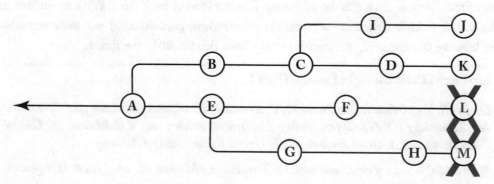

Figure 5.5: Flag A indicates a branching or pluralistic future. Flag B shows the continued progression of the threat future. Flag C indicated another branching (e.g., place). Various flags I and D show progression in different locations and differing threats J and K occur.

In a separate branch in Figure 5.5, Flag E shows the progression of the threat while Gate F prevents the threat. Similarly, in a different location flags G shows progression and gate H stops threat M from occurring.

The Goal of the TAD Backcasting

The ultimate goal of the TAD backcasting relates to the Threatcasting Foundation, specifically to the Application area. How will the results of the Threatcasting be used? What is the goal?

This phase is meant to take the EBM developed by the participant(s) and use the TAD backcasting to map out a comprehensive and actionable plan and set of flags for the ultimate consumers of the results. This phase is represented at the bottom of Figure 2.1 (Threatcasting Framework) as the participant(s) move backward from their imagined future. For figure simplicity, it is drawn as a single line, but now you understand the end product is more similar to Figures 5.3–5.5.

5.1.2   REVIEW THE EBM AND THE "EVENT"

Using their threat future (EBM) built in Phase 2, participant(s) put themselves back into their threat future and the Event. They then focus on counteracting the adversary and the negative future they have described. Here, once again, a diverse group of participant(s) becomes important. The steps that could be taken to disrupt, mitigate, or recover from the event can come from the whole of society. The interdisciplinary nature of the participant(s) means that they can pull from various perspectives (e.g., academia, government, industry, communities, etc.). Having participant(s) with a variety of experiences and expertise or who can pull from across these areas will expand the TAD across domains.

The detail of the EBM is important for the TAD Backcast. The person in a place experiencing the threat should have enough specificity so that participant(s) will understand both the point of view of their person as well as the adversary. This is a kind of back story. With an understanding for this future world's structure and multiple perspectives, participant(s) use their worksheet to explore how we disrupt, mitigate, and/or recover from this threat in the future.

Interview: Backcasting in Practice (Part 1)

Dr. Jon Brickey is Senior Vice President, Cybersecurity Evangelist, for Mastercard Operations and Technology (O&T). Before joining Mastercard, Brickey was a Colonel in the United States Army and Assistant Professor at the United States Military Academy.

With multiple years of experience with the Threatcasting Method, we asked Jon to offer general advice on Backcasting.

When you begin backcasting, in general the first step is to focus on the scenario and immerse yourself in the threat. Put yourself in the situation. It may require you to think of any situation you've been in yourself that was similar. Even though some of the threat futures can be pretty far out there and futuristic.

When you backcast, it also requires some knowledge of how the organization you are threatcasting for works. That's why it's good to have a diversity of thought and experiences in the room.

Backcasting is helpful because it helps people think about the order and sequencing of things that need to happen. In the military we called this backward planning. You start with the threat and the event, then you begin to lay out the sequence of events or things that have to happen in order to achieve a positive future or to disrupt a negative one.

You need to explore the preconditions that need to be in place and often doing that in reverse order can give people more clarity than if you started in the present and tried to think forward into the future.

5.1.3   GATES AND FLAGS

The goal of Phase 3 of the Threatcasting Method is to gather as much raw data from the participant(s) as possible. The facilitator should encourage them to fill in as much detail as they can. This detail will make the EBM richer and give the analyst(s) more to explore in the following phase.

Gates

Gates are actions that can be performed to disrupt, mitigate, and recover from the threat future. Who takes these actions and where they are located is as important as the action itself. Further detail should be gathered about who the principal actor is. Is it an individual or part of a group (government, military, private industry, academia, etc.)?

Generally, the initial gates the participant(s) develop are high level. The facilitator should encourage them to get as specific as possible. The gates will be used later in the TAD Backcasting as the basis and starting point for the milestones.

Example Gates include:

- government defender—regulations (or laws) that are passed, technological standards created, policies enacted, collaborations created, or resources spent.

- military defender—changes across the DOTMLPF-P (doctrine, organization, training, material, leadership, personnel, facilities, policy) spectrum for the workforce.

- industry defender—capabilities created, R&D efforts resourced, controls emplaced.

- academic defender—emphasis on academic research on a certain topic, change in educational structure and/or curriculum, ...

The participant(s) capture the raw data for the gates in the workbook (Figure 5.6).

PART FOUR– Backcasting - The Defenders (from the perspective of the defenders)		
Examine the combination of both the Experience Questions as well as the Enabling Questions. Explore what needs to happen to disrupt, mitigate, and recover from the threat in the future.		
What are the Gates?		
List out what the Defenders (government, law enforcement, industry, etc) do have control over to use to disrupt, mitigate and recover from the threat. These are things that will occur along the path from today to 2030.	1	
	2	
	3	
	4	
	5	

Figure 5.6: Example of Threatcasting Workbook focused on Gates.

Flags

Flags are events (e.g., economic, cultural, geo-political) or advances (e.g., technological, scientific) that defenders have no control over, but once they occur establish path dependencies with significant repercussions and consequence. Flags should have an irreversible effect on the envisioned future and should be watched for as heralds of the future to come. Types of broad flags could be technological developments, natural disasters, and economic or political catastrophe where once the event has occurred there is no going back; the event cannot be undone.

Flags are the indicators that a threat has begun to manifest itself over the decade span between the present. Flags are iterative, they build off one another. As in Figure 5.4, the first flag A enables the one to follow, flag B. They move from A to B to C. Flags do not move directly from A to D. In this case they would have been modeled incorrectly.

Typically, flags are derived from changes or advances in technology, science, and policy. They can also be driven from cultural or demographic shifts. Part 3 of Phase 2—Enabling Questions—is a precursor to the flags. The participant(s) are encouraged to imagine what needs to happen to enable or indicate that the threat is building. These flags can be large global events or local incidents.

Example Flags

- A new technology is developed that enables the threat

- A government regulation is put in place

- Global recession

- Significant natural disaster

- The adoption of a policy or technology begins to occur in a specific market or industry

- …

As participant(s) are exploring their flags, it can be helpful to prompt them for more detail. Once they have identified a single flag, ask what would happen before this flag to enable it to happen. For example, referring to Figure 5.4, perhaps the participant(s) have just described Flag C. What would be Flag A and B to allow Flag C to happen?

This detail and effort is important because it will generate early warning signs and signals to be monitored in the Application Area. Many find the flags helpful in business and security because it gives them an early indication that a threat might be arising. But it means they do not need to act too early or over reach. Well-articulated flags allow appropriate actions (gates) to be taken at the appropriate time.

Participants are asked to provide their flags for their threat future within the Threatcasting workbook, as seen in Figure 5.7.

What are the Flags?		
List out what the Defenders **don't** have control over to disrupt, mitigate and recover from the threat. These things should have a significant effect on the futures you have modeled. These are things we should be watching out for as heralds of the future to come. What are the technological/scientific advances that could be re-purposed? What are the incremental steps along the adversary's path?		
	1	
	2	
	3	
	4	
	5	

Figure 5.7: Example of Threatcasting Workbook focused on Flags.

The Duality of Gates and Flags

There exists a duality between the distinction of gates and flags. One individual or organization's gates could be a flag to another. It truly depends on the perspective. This is one of the interesting things that comes from the data in Phase 3 of the Threatcasting Method.

Interview: Backcasting in Practice (Part 2)

With multiple years of experience with the Threatcasting Method, we also asked Jon to offer general advice on Gates and Flags.

Gates and flags are essentially the highest level of detail in the backcasting. You explore what is within your control (gates) to disrupt, mitigate, and recover from the threat. The flags allow you to think about the preconditions or the enabler to the threat that are not in your control. To get as specific as possible can require some domain expertise. Meaning, you will need people who understand the area in which you are threatcasting. If you are looking at threats to farming it would be good to have a farmer because they are going to know the details. But it would also help to have a technologist who understands how technology works. This will broaden the perspective and possibilities. If you have three people sitting around the table that all have similar background and experience, then it's tough to get a new perspective.

One way we think about flags are as specific things to keep an eye out for. Once we have identified a range of threat futures and done the backcasting, the flags give us the indicator that the threat is beginning to happen. When a flag does happen, then you know as an organization that you need to take action. That action is a gate, and you know what to do. You have a playbook because of the backcasting exercise. You have a place to start.

When you think about creating that playbook, again people with different backgrounds are important. For example, in cyber security where I work, people generally think about technology and security. But when you backcast it's important to get the opinion of legal or someone else in human resources. They will bring up something you never thought about. All threats span multiple domains, experiences, and people.

Also, the 10-year timeframe is helpful for gates. Looking out 10 years means that as an organization you have time to take the action, time to make the change. To effectively disrupt some threats, you might need to make a large change. Like changing your entire technical infrastructure. The decade timeline means you can get really specific about the steps you and the organization need to take over time.

FLAGS

A flag is an indicator. It is like an official waving a flag during an auto race. The flag indicates something to the observer. If a yellow flag is displayed, the drivers know the race is under caution. When a checkered flag is waved the drivers know the race is complete.

Generally, when a flag occurs there is no going back from it (e.g., the invention of a new technology, the appearance of a new kind of business, the filing of a patent), or at least it would be difficult and time consuming (e.g., the passing of a law or regulation by a government).

GATES

A gate is an action that can be taken by the "Who" that has been outlined in the Application Areas of the Threatcasting Foundation. These actions are recommendations to be taken from the participant(s) in the workshop and ultimately the analyst(s) in the Findings.

These actions can be quite varied from high level conversations to investments or physical actions.

Examples:

• Fund a specific kind of research.

• Convene people together to discuss a topic or plan a response.

• Share information across a wider set of organizations.

• Build a specific product to solve a problem.

• Pass a law or regulation.

• Invest in a specific area, population, or market.

• Take a specific physical action.

An Explainer of Incrementalism

Most people use Incrementalism without ever naming it because it's a natural way to complete daily tasks such as getting dressed. Logical Incrementalism implies that the steps in the process are sensible (e.g., putting on socks before shoes).

When it comes to organizations, logical incrementalist Charles Lindblom (1959) asserted that most organizations are deeply built upon their past actions and use them to identify their future course of action.

Using Incrementalism as a tool is useful because it's practical, it's responsive to uncertainty and the reality of logic, it is well suited for extreme complexity, it adapts to handle psychological shifts, and it can be used to build a solid shareable plan.

Incrementalism was first developed in the 1950s by the American political scientist Charles E. Lindblom in response to the then-prevalent conception of policymaking as a rational analysis process culminating in a value-maximizing decision ("Incrementalism", 2021).

5.1.4 MILESTONES

Milestones, as defined by *Merriam-Webster Dictionary*, are significant points in development. Using the defined gates as a starting point, participant(s) now place them on a developmental timeline. Then, using the TAD backcasting approach, these milestones can be expanded to include physical location and specific people involved in the action. The timeline is critical because it forces the participant(s) to see these actions' iterative nature and early impacts. Therefore, as the threat is a decade in the future, the people and organizations performing the actions to disrupt, mitigate, and recover have time to monitor and act.

There is no magical formula on how to cut the decade into accomplishable sections. The default is four- and eight-year timeframes. Ten years is just far enough that to disrupt it, specific actions will need to happen eight years from the present (that is, two years before the threat/event manifests). To enable the eight-year out milestones, the time is cut in half and the Practitioner(s) are asked to imagine what needs to happen four years from now that would enable the actions eight years in the future. With most organizations, anything significant that needs to be completed four years in the future will have to start in the very near future.

The TAD backcasting creates a matrix of actions and indicators, that then spawn ideas about other actions to be taken. It is quite complicated, pluralistic, and nested. Typically, Practitioner(s) do not build their milestones linearly. Instead, it is more typical that they will move up and down the timeline, filling in the interactive steps that need to be taken, along with who and where they need to be taken.

Participant(s) fill out the milestones in the Threatcasting workbook, as seen in Figure 5.8.

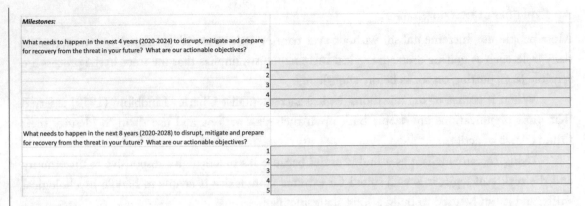

Figure 5.8: Example of Threatcasting Workbook focused on Milestones.

It is often difficult for Practitioner(s) to complete the entire TAD Backcasting. This is normal, especially during the first round of Threatcasting. When the participant(s) are new to the Threatcasting Method there will be a learning curve. Optimally the analyst(s) should try to facilitate two to three rounds of Phases 2 and 3 by the participant(s). This will give the analyst(s) a larger volume of raw data and threat futures to be analyzed in Phase 4.

Interview: Backcasting in Practice (Part 3)

In the finale of the three-part interview with Jon, we asked him to offer general advice on milestones.

Milestones are the next level of details after the gates and flags. They are essentially the gates, but with much more detail laid out over time.

For example, if you're trying to do something in four or eight years that requires a change in a law or regulation, you know that this doesn't happen right away. There are a lot of steps that need to happen to get a law changed or created. By backcasting the milestones, you can lay out intermediate steps and goals that build off each other. If you are a company and you want to change a law, then you might want to start working with a lobbying company that can help influence lawmakers. Or, you might want to build that lobbying capability inside of your enterprise. But somehow you've got to reach that goal way before the law will get changed. Before you hire a company or grow your organization, you will probably want to examine the steps you'll need to take for each option. What will it cost? How long will it take? Next, you'll want to reach out to people and set up contacts to get more information and grow your network.

The milestones really show that there are a lot of small actions that you'll need to take to accomplish your future goal. But you have to be specific, break it apart and understand what step enables the next step. You also don't have to have all the answers at the beginning. Part of the process allows you to uncover key decision points, and then your backcasting can branch into multiple possibilities. This is how the Threatcasting Method is different from most planning. It allows for multiple futures so you can prepare for multiple possible futures. As a group you throw out possibilities and capture them. It's structured spitballing!

5.1.5   WRAP-UP

One key benefit and output of the Threatcasting Method is its exploration of potential second- and third-order effects of these actions within the future. This is especially useful for large and complex military and business organizations. The generated SFPs craft an easy and quick way to understand the visions of the future, giving these organizations a quick way to understand threats and discuss what actions need to be taken. "People aren't wired to imagine the future, ten or even five years out, which is a blocker to innovation," said Kate O'Keeffe, senior director of Cisco's Hyper-Innovation Living Labs (CHILL) who used Threatcasting in 2017 to explore future threats to the digital supply chain. "We need to create that world for them, so they can immerse themselves in this future scenario, making it immediately apparent what kind of solutions we need to prepare for that future" (Johnson and Vanatta, 2017).

At the end of Phase 3, the participant(s) report out, telling the larger group a story about their SFP. They describe the envisioned threat and then work backward to explain what could be done to disrupt, mitigate, and recover from that threat. Threatcasting Phases 2 and 3 are traditionally repeated between three and four times with the group of participants. This allows people to become comfortable with the Threatcasting Method, the workbook materials, and the moderators' questioning. For each iteration, the participants choose new Foundational data points from the research prompts. Familiarity with the process and materials free the participant(s) to develop a broader range of detailed futures to explore a wider range of possible threats. Additionally, the moderators take more time, pushing the groups to expand their SFPs to expose more of the possibility space.

At the end of each round of Futurecasting and Backcasting, the working group of participant(s) will report out their work in a limited time. This allows all of the participant(s) to hear and be affected by each smaller working group's threat future. Note: The logistics of this will be explored in Part 2 of the book.

Applied Futures Note

Between Phases 3 and 4 is a significant amount of work taking the raw data from the threatcasting workbooks and preparing it for the post analysis and synthesis. As this significantly differs depending upon the type of Threatcasting (individual to small group to large group), we will explore these steps in Part 2 of this book.

5.2   POETRY

Star Gazer

Forty-two years ago (to me if to no one else
The number is of some interest) it was a brilliant starry night [9]

Louis MacNeice

Why This Poem Matters

MacNeice's poem is a meditation not just on light traveling from stars but of time on a galactic scale. In the poem he remembers back 42 years when he gazed up at stars, knowing vaguely that the light from those stars would take years to reach his and that he in fact was looking at them from the stars left before he was even born.

Even though when we backcast using Threatcasting we may look ahead just ten years, a mere blip in time for MacNeice's poem, but it is important to remember the scale and span of time. To stretch ourselves to see over these vast time scales so that we might better place ourselves in them.

5.3   EXERCISES

The best exercise for the analyst(s) to practice backcasting is to apply the Threatcasting Method in practice. We have provided a mock threatcasting in the Project portion of this book to assist. The following three exercises are skill-building.

5.1

This exercise explores the concept of incrementalism.

1. Think about a task that can be completed in five steps or fewer. Examples could be putting on socks or pouring a glass of water.

    a. Write the steps down.

    b. What would happen if step #3 and step #4 were reversed?

    c. Does it matter if these two steps were reversed to accomplishing the task?

2. Think about a task that has more than ten steps in order to complete. Examples could be making a sandwich or booking a flight.

---

[9]   Read the full text at https://bit.ly/3kS25Pm.

    a. Write the steps down.

    b. Is it possible to sabotage this task by switching steps? If so, which steps?

3. Think about a project that has so many steps a digital tool would be needed to log the number of steps. Examples: Landing a Mars Rover, Time travel, solving climate change.

    a. Write down an extremely complex multi-step tool or project of your own.

    b. What are some opportunities to sabotage this project or tool in its development?

5.2

Gates and flags function on a time horizon. It's important to know how to build a timeline to look ten years in the future. A timeline is also a map to follow for what will happen at the two-, four-, six-, and eight-year markers. Use actual dates on your timeline as it helps to imagine the future. This time horizon asks you to disrupt, mitigate, and recover from the event.

1. Start by thinking of a possible negative event ten years from today.

2. Write down a detailed description of your event.

3. Write down who specifically this future event affects. For the "Who," consider individuals, populations, communities, organizations.

4. Write down who this would be a negative future for.

5. Write down who this would be a positive future for.

6. Working backward from ten years in the future to the present time, write down:

    a. How could you disrupt the event from occurring?

    b. How could you mitigate the event (i.e., reduce the impact to the communities for whom this is a negative future)?

    c. How could you recover from the event?

7. Write down (at the two-, four-, six-, and eight-year markers on your timeline) what kinds of flags and gates might appear that could be used for indicators to take action.

5.3

Enabling the Enabler

In this exercise, you will explore building blocks or how to enable the Enablers.

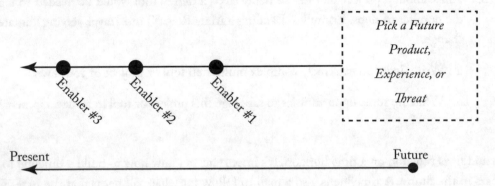

- Pick a product, experience, or threat in the future (you can use previous examples in this book).

- Enabler #1

   ○ Begin Backcasting by asking: What was a technology or investment that enabled the product, experience, or threat to happen?

- Enabler #2

   ○ Think about an investment, technology, group of people, or research that would have enabled Enabler #1 to happen.

- Enabler #3

   ○ Finally, what investment, technology, group of people, research, or event (e.g., conference, meeting, etc.) would have enabled Enabler #2 to happen?

By breaking down and exploring each enabler that enables the follow-on enabler, the analyst(s) begins to see that all gates, flags, or milestones are enabled by a previous enabler. The analyst(s) can continue this exercise for as long as it is productive for the application areas.

The takeaway from this exercise is to know it is possible and necessary to understand and document what enabled each enabler to give the "Who" in the Threatcasting Application Area an actionable level of detail to either plan the incremental steps (gates) to disrupt, mitigate, and recover from a potential threat, or provide a high level of enabling details for a flag so that it can be identified early when it begins to manifest.

CHAPTER 6

Phase 4

"If we do not learn from history, we shall be compelled to relive it. True. But if we do not change the future, we shall be compelled to endure it. And that could be worse."
—Alvin Toffler

6.1 POST ANALYSIS, SYNTHESIS, AND FINDINGS

After completing the workshop and generating the raw data, the next phase in the Threatcasting Method is post analysis where the data is synthesized and ultimately a set of findings are developed. In this phase, analyst(s) review and process the raw data to generate a set of findings (possible and potential futures). These findings, along with the specific actions (gates) and indicators (flags) will be used to generate a report out. The various types and potential methods of delivery for this "report out" will be covered in Chapter 7.

The activities within this phase are:

- prepare the raw data for analysis,

- preform multiple rounds of analysis on the raw data,

- penerate, validate and review findings, and

- specify actions to be taken and indicators to be monitored.

6.1.1 THE ROLE AND ART OF THE ANALYST(S)

In Phase 4, the analyst(s) review and process the raw data. Depending upon the size of the raw data set, scope of the Threatcasting, and setting (e.g., corporate, military, academia), the number of analyst(s) will vary. Although it is not a requirement, having multiple analyst(s) can be helpful for collaboration, debate, and validation.

Each analyst(s) brings their own expertise, background, experience, and biases to the post analysis. Just as the Threatcasting Method recognizes, embraces, and curates the expertise, background, experience, and biases of the workshop participant(s), the analyst(s) role is approached in the same way.

To use a different mental model, for the analyst(s) the Threatcasting Method, framework, and process is like a piano. If the instrument is the same, it is standardized. Additionally, the notes and the music used to inform the playing of the instrument are standardized as well. However, every person plays the piano in a different way. Each pianist has a different style or approach.

Understanding and often stating the analyst(s)' expertise, background, experience, and biases is important. Each analyst(s) will also have a specific relationship to the research question and threat areas being explored. When multiple analyst(s) are working together during the post analysis, it can be helpful to seek out and curate different analyst(s) for the team. Each will bring a different perspective, and this diversity will produce stronger findings.

For example, if a team was Threatcasting future threats at the intersection of banking and connected technologies, having analyst(s) or a single analyst with experience and expertise in these areas will be advantageous. But bringing in an analyst(s) that has experience with underserved communities or a specific generation of consumers could also be helpful. The different perspectives of the analyst(s) and the collaboration between the team could illuminate new threats and patterns in the raw data and findings.

6.1.2   PREPARE THE RAW DATA SETS FOR ANALYSIS

Directly after the Threatcasting Workshop, the raw data sets and prompts need to be prepared for post analysis. This preparation includes data hygiene and anonymization. These actions will ensure that the data sets are uniform and easier to process by the analyst(s). Example actions can include: cleaning up any participant(s) formatting changes, placing data into the correct cell, and general layout of the data set.

Additionally, assuring the raw data sets are properly managed and stored will allow them to be used by other analyst(s) in later Threatcasting Projects. When the Threatcasting Method is consistently applied to multiple Threatcasting Foundations, data sets from other workshops can be used and input for different projects where appropriate.

Data Hygiene

Data Hygiene is a term used to describe a process by which analyst(s) will "clean" the data so that it will be the most efficient and usable for the analyst(s) as they perform post analysis and synthesis.

Data Hygiene steps to take include:

- standardize data;

- validate data;

- de-duplicate data;

- analyze data quality; and

- investigate and repair any quality problems.

The workbooks are a structured database (e.g., spreadsheet). They are in a standardized form and format so that all of the raw data is similarly structured from all the participant(s). However, there are always outliers that will require manual modifications. This will make the raw data easier for the analyst(s) or software programs to analyze, depending on the tools the analyst(s) choose. Making sure that all the data is structured the same is important.

The next step will be to validate that the participant(s) have entered all the data into the workbooks. If there is data missing, the analyst(s) will need to decide if they need to follow up with the participant(s) to fill in the missing data or if it is acceptable to leave the data missing. Cleaning up the raw data will also include searching for duplicate or repeated sections in the data.

After these first three steps have been taken, the analyst(s) can then analyze the data sets' quality to make sure there is enough data and that it is in a form that is ready to be analyzed in the post-analysis phase.

During this review, analyst(s) can also use this time to investigate any possible problems that might inhibit the post analysis. This will be the phase in the methodology to repair, adjust, or take further actions to make sure the data is complete and ready.

Anonymization

During the post analysis it will be helpful for the analyst(s) to keep the participant(s) identities tied to the specific parts of the raw data they generated. The analyst(s) may need to follow up with questions if the data is not clear or ask for clarification if the data is not complete.

However, in preparation for the findings and report out, it is important that the data be anonymized, meaning that when the report out is delivered, the reader or consumer of the report will not be able to connect the participant(s) with any specific piece of raw data. Aside from allowing each participant(s) to feel comfortable to generate any threat future they can envision, it means that the raw data generated does not contain Personally Identifiable Information (PII).

PII is a unique piece of data or indicator that can be used to identify, locate, or contact a specific individual. Permitting identification or recognition of a unique person is what distinguishes PII from other types of personal information (information that may be sensitive, embarrassing, or offensive that an individual may wish to control, keep private, or not have disclosed to others without consent) (Staples, 2007).

Interview

Jason Brown has spent over 20 years with the U.S. Army in leadership positions within cybersecurity, information warfare, intelligence collection, and analysis of threat groups during multiple deployments to the Middle East and Afghanistan. Currently he is getting his Ph.D. at Arizona State University in the School for the Future of Innovation in Society and the Threatcasting lab.

We asked Jason to comment on his experience with the Threatcasting Method's post-analysis phase and any advice for analysts.

The Importance of the Application Areas

The application areas in your Threatcasting Foundation are so important. They will not only keep you on track, but they will help answer the research question. Because just answering the research question isn't good enough. The analyst needs to answer the research question for someone. And that someone needs to know not only why this information is important but also what should be done about it.

My years of experience in the military really helped with the post analysis. For years I was the person in military intelligence and operations who had to give reports to operations advisors and commanders. When you do that, you have to give them the essence of the information, fact or intelligence but then you have to follow it up right away with why it's interesting and what do—Soldiers need to be armed with that so they can take action right away.

In Military Intelligence there is something called this commander's priority intelligence requirements. It's often a question. Something like: What's the location of the enemy fuel suppliers?

The reason the commander is asking the question is not because they want to know where the enemy fuel supply is. The real reason they are asking is because they want to disrupt the enemy's fuel supply.

That's a good way to think about the relationship between the research question and the application areas. All the indicators that you pull out of the Threatcasting raw data need to both answer the research question, but ultimately serve the end goal as well.

The Importance of a Sticky Note

When you start the post-analysis phase, the raw data is like a big jar full of information. You need to categorize it and get to know the topic. The research question can provide you a filter to run all of the information through. You can ask yourself, does this answer the research question? Does it shine new light on it or give new perspectives to the question or topic?

Often, during the post-analysis phase, I will write the research question on a sticky note and put it on my monitor. I keep looking back at it to see if the data I'm analyzing and the categories, patterns, and clusters that are starting to show up in the data answer the question. These become focusing events for me and help me narrow down the data.

6.1.3   PERFORM MULTIPLE ROUNDS OF ANALYSIS ON THE RAW DATA

A Note on Post-Analysis Tools

There is emerging work in the Threatcasting Method that is exploring the use of multiple procedures and tools (both cognitive and digital) to aid analyst(s) in the post-analysis phase. For the sake of this chapter, we will outline the tasks and desired outcomes from each step or round in the phase. There are three rounds of analysis that are typically done on the raw data in order to determine the Threatcasting findings. Part of the art of the analyst(s) will be to choose what process or tool best suits them, the team, and the research question.

Note:

All the work of the analyst(s) is captured in a post-analysis workbook. This provides a shared working space if there is more than one analyst. Also capturing each step of the post-analysis process ensures complete transparency in the process and aids in writing the final report.

Rounds 1 and 2 of post analysis use a technique called Axial Coding.

> *"Axial coding is the process of relating categories to their subcategories. Anselm Strauss and Juliet Corbin used this term in Basics of Qualitative Research as one of the data analysis techniques by which grounded theory can be performed. The essence of axial coding is to identify some central characteristic or phenomenon (the axis) around which differences in properties or dimensions exist. Axial coding is therefore a process of reassembling or disaggregating data in a way that draws attention to the relationships between and within categories"* (Wicks, 2010).

Round 1: Summary

The goal of Round 1 of post analysis is to derive a summary of the raw data, separating each threat future the participant(s) generated in the EBM. This will make the raw data more manageable and also give the analyst(s) a first overview of the threat futures. Later, once the findings have been derived, the analyst(s) will reexamine the EBM to pull out the gates and flags (actions and indicators).

For Round 1, analyst(s) use the Futurecasting portion (Parts 1, 2, and 3) of each EBM. The participant(s) were encouraged to envision and imagine a broad range of threat futures, pulling together a large EBM. The summary section tasks the analyst(s) to summarize the most important points covered in the raw data as related to the research question. The key to a successful summarizing round is to avoid interpretation at this stage. Be true to the intent of the workshop participant(s) and be ruthless in using their words as much as possible.

If Round 1 is successful, then the analyst(s) will rarely need to return to the full data set unless they are investigating a specific detail or searching for an anomaly. Round 1 also makes the data set a more manageable size to continue to the next steps.

There are two primary techniques the analyst(s) use during Round 1:

- Length reduction

- Summarize

Length Reduction

The first goal of Round 1 is to reduce the length of the EBM from multiple pages of raw data to a single text block. Typically, these text blocks are one to two paragraphs. The length of the paragraphs should be sufficient enough to capture parts 1–3 of the Futurecasting sections of the EBM.

The analyst(s) can leave out most details, however if a particular detail appears important, it should be left in.

Summarize

In Round 1, the analyst(s) attempts to reduce the word count for the raw data while still capturing the essence of what the participant(s) imagined in the EBM.

A summary is "a brief overview of an entire discussion or argument. People often summarize when the original material is long, or to emphasize key facts or points. Summaries leave out detail or examples that may distract the reader from the most important information, and they simplify complex arguments, grammar, and vocabulary" (MindTools, n.d.).

An example process for summarization:

1. Get a general idea of the original.

   First, speed read the text that you're summarizing to get a general impression of its content. Pay particular attention to the title, introduction, conclusion, and headings and subheadings.

2. Check your understanding.

   Build your comprehension of the text by reading it again more carefully. Check that your initial interpretation of the content was correct.

3. Make notes.

   Take notes on what you're reading or listening to. Use bullet points and introduce each bullet with a key word or idea. Write down only one point or idea for each bullet. Make sure your notes are concise, well-ordered, and include only the points that really matter. Tip: The Cornell Note-Taking System (Basso and McCoy, 1996) is an effective way to organize your notes as you write them, so that you can easily identify key points and actions later.

4. Write your summary.

Some summaries, such as research paper abstracts, press releases, and marketing copy, require continuous prose. If this is the case, write your summary as a paragraph, turning each bullet point into a full sentence.

Aim to use only your own notes and refer to original documents or recordings only if you really need to. This helps ensure you use your own words. If you're summarizing speech, do so as soon as possible after the even, while it's still fresh in your mind.

5. Check your work.

Your summary should be a brief but informative outline of the original. Check that you've expressed all of the most important points in your own words, and that you've left out any unnecessary detail.

Therefore, you might need to read through the EBM several times and record a true summary of the scenario that tells a cohesive story. When appropriate, include the actor's name and specific details about how they experience the event and how the threat came to be. The benefit of this process is that the summary becomes more readable in subsequent rounds and you, as an analyst, become immersed in the scenario. The drawback is that details are smoothed out and may become lost. It may require you to refer back to the raw data workbook several times in each round to ensure you are being true to the intent of the workshop participant(s).

Figure 6.1 shows a sample post-analysis workbook after Round 1.

B	C	D
Round	**Team**	**Round One - "Summary"**
1 Red Pawn		A young mother in the first nations community works in shipping transport at her local Walmart in Paducah, KY - the 3rd most prominent micropolitan area in the US. A Chinese multinational company wants to capture marketshare in the US via the largest retailer network. Their first goal is to create a dependency on their own network by interrupting and rerouting all Walmart empoyee data and systems traffic. They create social media narratives - recycled memes, old stories about shabby product, underpayment of workers, AI autogenerated communities of fake workers complaining - in first the native american, then other subgroups, including the head of logistics, to sow seeds of doubt and unrest across Walmart employees. Second stage of the attack includes phishing emails and messages disguised as a link to the AI genreated Audio/video deep fake "exposee" videos - to secure logisitcs/operations credentials from workers and employees - our young mother and every other transport tech have smart headsets with customer data and unique explicit user profiles/shopper hisotry. Through targeting her access credentials, the malactor can get into the Walmart network, Walmart's defenses work on a corporate level, not an individual actor working
1 Orange Pawn		Joe Snuffy, a social media influencer at a University, living in a low economic suburb of St. Louis. An extremist/hactivist group is looking to disseminate their message and target him as a catalyst for dissemination, leveraging the legitimacy of his podcast and extensive following. Hactivists create deepfakes of Joe's podcast, recreating his voice perfectly, and swaying his subject matter just enough to continue to capture his regular audience with their own targeted message. Listeners begin questioning his views, peers disassociate, the university questions his motives causing panic.
1 Yellow Pawn		Akito is the daughter of the Japanese Emipirical throne reestablishing the Empire and dominance of Japanese culture through the Pacific. She's responsible for establishing the resurgent warrior culture in Japanese culture. Japan is mobilizing their community to regain their regional dominance that they had prior to WWII. As climate change impacts their landmass, they are looking for new territory to preserve the cultural /genetic empire. Akito is a target of Chinese IW trying to divide her (as an influential political and military symbol) away from achieving Japan's "first boots on land" strategy. Finding deepfake videos of her 7 year old daughter in compromsing sex acts, Akito is forced to reprioritize her
1 Teal Pawn		Mrs. Foley is a middle school civics teacher in Iowa. As urban populations increase nation-wide, rural communities have less influence on democratic outcomes. Voting mechanisms have broken down (vote via pop up ad) which has caused lack or representation and increased frustration particularly in rural areas. The anti-democratic party sees this as an opportunity to influence in the most vulnerable regions. In 2029
1 Purple Pawn		Corporations (information tribes) establish ownership over information - they are the controllers and governance of all information for purpose of market dominance. Throught this information dominance, major cultural divides emerge between information tribes, as well as between urban and rural populations. Those living closer to information hubs have access to virtual goods and benefits (apps, social media, messaging,

Figure 6.1: Example of Post-Analysis Workbook after Round 1.

Round 2: Meaning

Using the output from Round 1, the analyst(s) is tasked to review each summarized threat future and capture the meaning of the threat future in context to the Threatcasting Foundation (topic, research question, and application areas), the prompts and any additional information from discussions that took place during generation of the raw data.

The analyst(s) can explore the following questions.

- How does the summary address the topic?

- How does the summary answer the research question?

- Does the threat future provide a new perspective on the foundation?

- What is the significance of the threat future to the research foundation?

- What is essential and what can be left out?

In Round 2, the analyst(s) begin to interpret the data. Analyst(s) look for interesting themes within each scenario or that span across multiple scenarios. Analyst(s)' measure of "interesting" is somewhat arbitrary, but generally it fills one of two criteria: either they are unique to a scenario and stand out as an outlier, or they link with several other scenarios and could lead to a finding.

Round 2 will greatly reduce the detail from Round 1, identifying clusters and patterns from the raw data. During Round 2, the summary is reduced to its high level meaning with shorthand notes for possible clusters or patterns (see cross round techniques discussed later in this chapter).

There are two primary techniques the analyst(s) use during Round 2:

- Broader connotation

- Prompts comparison

Broader Connotation

Using the summaries from Round 1, the analyst(s) compares the threat futures against the research foundation to understand what connotation they might have. In other words, what does the summary of the threat future mean to the topic, research question, and application areas?

The analyst(s) can explore the following questions.

- Does it uncover new implications?

- Does the threat future have a broader meaning to the foundation beyond the threat future itself?

- Are there any topic areas that stand out as special?

- Are there broader actions at work?

Prompts Comparison

As the name indicates, analyst(s) will compare the summaries of the threat futures to the information in the prompts. Considering all of the prompts, the analyst(s) can explore the following questions.

- What topics, areas, or issues does the threat future address?

- Is there a single theme that stands out?

- Is there a contradiction to the prompts?

- Is there a unique detail that was surprising?

When thinking about how to record the meaning in the EBMs, there are several techniques to code the ideas identified in the data into your post-analysis workbook. The first is to use the words of the scenario itself to describe the effects of the model. This is called in vivo coding. Another way to think about these patterns is to describe them in gerunds (nouns turned into "-ing" verbs). Generally, short phrases and bullet statements are usually appropriate for finding meaning, because the emphasis is on simplifying complex ideas into simple building blocks ("divergent thinking") before recombining them into larger clusters later ("convergent thinking").

Figure 6.2 shows a sample post-analysis workbook after Round 2.

Round	Team	Round Two - "Meaning" / "Insight"
1	Red Pawn	system exploitation on two levels - data network (data systems, credentialing, logistica) and social network (emotional belief networks of employees, particularly those populations already feeling struggle) - creating synchornized doubt and vulenrability both soft and hard for a more pervasive attack.
1	Orange Pawn	anyone can be exploited. It's no longer those just with extreme/notable power or influence, but more and more those who simply have an audience (mid-fluencers)
1	Yellow Pawn	priorities will always lie with cultural /social undercurrents - those things that are generations old and built into culture. Targeting vulnerabilities based on those cultural / social undercurrents will be exponentially more effective for reflexive control
1	Teal Pawn	information capitalism is the next iteration of disinformation dissemination / 'fake news'. Targeting moves from established decision makers to the emerging decision class - hyjacking
1	Purple Pawn	information access divide based on data capitalist practices. The same have/have not trope has shifted into an information access gap, exacerbating the instability shared information for
1	Black Pawn	control of pervasive non-critical data infrastructure (educational algorithms, textbooks, hiring practices) to behaviorally manipulate toward a single desired outcome. Those systems that are not top secret are now the most vulnerable and most

Figure 6.2: Example of Post-Analysis Workbook after Round 2.

Round 3: Novelty

Round 3 is the final round in the post analysis. Using the output from Rounds 1 and 2, the analyst(s) explores what is novel, or what stands out from the previous rounds. This novelty should directly answer or give a new perspective to the research foundation.

This is where the art of the analyst(s) is applied.

- What is the data telling you?

- What was meant?

- Are new threats or issues uncovered in context to the research foundation?

- How is the research question answered in a way that has not been addressed previously?

- What threat futures are new in the range of the possible and probable?

- What information might be useful to take specific actions in the application areas?

- Are there any clusters, patterns, or outliers?

Transforming these lists of themes (from Round 2) into distinctive categories that illustrate the "novelty" of a particular future or group of futures comprises Round 3. Rapidly moving from the individual themes in Round 2 to novel findings in Round 3 requires both disciplined thinking and experiential points of view to answer the "so what?" of the matter.

A novel finding can be a new concept to the organization or as a new instance of something seen elsewhere. Another way to approach novelty is a concept that will have profound implications to the application area. This kind of novelty is usually unprecedented and may not correlate or transfer to other situations.

Across all three rounds there are techniques available to the analyst(s) to help in the identification of novelty. Consider these to be cross-round techniques.

- Clustering

- Reflection

- Outliers

Clustering

During each round the analyst(s) looks for clusters and patterns in the data. Clusters and patterns can take on nearly any form including terms, words, themes, futures, characters, technologies, details, etc. The analyst(s) looks for any commonality in recurring threat actors, technologies, places, etc. that might appear in multiple EBMs.

Throughout the post analyst, the analyst(s) notes these clusters, calling them out in the analyst workbook.

Reflection

Once the analyst(s) has identified possible clusters, they reflect on why the clusters are there and whether they provide new perspectives or answers to the Threatcasting Foundation.

Outliers

Across all of the rounds—but primarily in Round 3—the analyst(s) begins to consider the outliers.

- Is there a single threat, actor, or technology that stands out?

- Does this single threat or aspect inform or serve the Threatcasting Foundation in a new and meaningful way?

- Does the outlier contradict or conflict with other futures?

What was not discussed?

Based on a review of the prompts, the analyst(s) examines what could have been discussed in the raw data but was not.

- Are there areas for more exploration?

- Is there a blind spot or bias in the data?

- Was there a subject that wasn't addressed?

    ○ Specific group of people

    ○ Technology

    ○ Location

6.1.4   GENERATE, VALIDATE, AND REVIEW FINDINGS

The clusters and patterns identified in the three rounds of analysis provide the basis for the findings. The findings are how the clusters and patterns answer the research question and give new perspectives to the entire Threatcasting Foundation. They can identify specific possible and potential threats or a broad threat area.

The validation of the findings can take several forms. One way is through collaboration and sharing between multiple analyst(s) if the work is being done by a team. Often analyst(s) will conduct the three rounds independently and then convene to compare findings. This collaboration can be helpful to refine the final findings but also expose new areas that might have been identified by only a single analyst (or show explicit or implicit bias).

Another form of validation is to have any SMEs that might have been used as prompts to review the findings. As their research and work were inputs to the Threatcasting, their perspective and validation of the findings can be helpful. Additionally, analyst(s) can return to all of the participant(s) who generated the raw data and validate the findings with them. This is a form of group peer review where the analyst(s)' interpretation of the raw data is explored. This also allows all participant(s) to comment and make sure that their perspective and opinions were captured correctly.

6.1.5   SPECIFY ACTIONS TO BE TAKEN AND INDICATORS TO BE MONITORED

Once the Findings have been generated and validated to the analyst(s)' satisfaction, the analyst(s) return to the raw data as well as to the post-analysis workbook to identify the specific actions (gates) that can be taken to disrupt, mitigate, and recover from the threats captured in the findings. The analyst(s) also identify the indicators (flag) that relate to the findings that will show the threat futures identified are beginning to materialize.

Analyst(s) explore how the various EBMs have been clustered, combined, or nested together. The analyst(s) review the Backcasting for all EBMs and capture the gates, flags, and milestones that apply to each of the Findings.

To expand the actions and indicators, analyst(s) can refer to the prompts, conduct further research, and/or conduct follow up interviews with SMEs or participant(s). This will ensure that the findings are expanded to answer the application areas.

6.2 POETRY

The Peace of Wild Things

When despair grows in me
and I wake in the night at the least sound [10]

Wendell Berry

Why This Poem Matters

Post analysis can be daunting. Especially as the lead analyst you will find yourself "neck deep in data," and we can sometimes feel like we are underwater. Futures, threats, clusters, outliers—as an analyst you can get lost in all of the information.

Berry's poem reminds us to take a break. Go for a walk. Get away from the work. If you take Berry's advice literally (and you may want to), go for a walk in nature. Don't think about the post analysis. Don't think about the threats. Don't think about the future. "Rest in the grace of the world" for a little bit. Clear your head. The post analysis will wait and when you return, you will be all the better for it.

---

[10] Read the full text at https://bit.ly/3kRpNuO.

CHAPTER 7

Phase 5

"The value of an idea lies in the using of it."
—Thomas A. Edison

7.1 OUTPUT

In this chapter, we will look at high-level, best-known methods for capturing the findings from the Threatcasting Method to enable analyst(s) to best communicate with potential audiences. Every individual and organization expects to receive information in different ways or in different formats.

The activities within this phase are:

- review traditional methods for delivery of Findings;

- review alternative methods for delivery of Findings; and

- determine the best combination of methods.

7.1.1 HOW TO DELIVER FINDINGS

Because Threatcasting is an applied Futures Method, how the findings are applied is important to the final phase. As analyst(s) begin to generate the output, they should be thinking about who is going to use it and how they're going to use it. Ultimately, the results analyst(s) get from Threatcasting only matter if they can package the findings into a narrative that others will find useful to empower their actions and strategy.

7.1.2 TRADITIONAL METHODS FOR DELIVERING FINDINGS

The Technical Report

The technical report is one example of how to package the findings from your Threatcasting Workshop. Technical reports (also known as scientific reports) will describe the Threatcasting Method, provide the Threatcasting Foundation (topic and research question), and showcase the results and findings from the analyst(s)' post analysis as well as include the raw data.

A key attribute of the Technical Report is transparency. The ability for the reader to quickly understand the inputs and research for the Threatcasting Workshop, have access to the raw data, and comprehend the findings will aid with a quicker acceptance of the findings, indicators, and proposed actions.

Technical Reports typically do not go through a traditional academic or journal peer review. However, often a peer review can be done with both the workshop participant(s) and the SMEs. This allows for some oversight of the findings. It also gives each reviewer an opportunity to add or modify the report. Additionally, it provides the opportunity for the analyst(s) to go back to the participant(s) and SMEs with any questions or requests for more information.

Sample Outline/Table of Contents

- Executive Summary

- Overview of Threatcasting

- Introduction

- Threatcasting Workshop Goals (Threatcasting Foundation)

- Threats

- Flags

- Actions

- Conclusion

- Appendices with all the Raw Data

For the Technical Report, a best-known method for the presentation of interesting or novel threat futures is to include short vignettes interspersed throughout the body of the report. These vignettes can be thought of as short teasers for the SFPs the participant(s) generate. They are like movie trailers that hint at the threat and help illustrate the findings in the report. It can make the technical report less dry and give the reader a feeling for what different threat futures might actually be like if they happened.

The final report is usually a text document (Microsoft Word or PDF) and the layout can include pictures and illustrations to further support the findings.

Resources

The following are examples of technical reports written from a Threatcasting activity.

- The New Dogs of War: The Future of Weaponized Artificial Intelligence—https:// threatcasting.asu.edu/publication/threatcasting-report-future-weaponized-artifi- cial-intelligence.

- Information Disorder Machines: Weaponizing Narrative and the Future of the United States of America—https://threatcasting.asu.edu/publication/threatcasting-report-information-disorder-machines.

- Information Warfare and the Future of Conflict—https://threatcasting.asu.edu/publication/threatcasting-report-information-warfare-and-future-conflict.

- The Regional Cyber Futures Initiative: The Future of Risk Security and the Law—Threatcasting Report 2019—https://uploads.prod01.sydney.platformos.com/instances/157/assets/images/The%20Future%20of%20Risk%20Security%20and%20the%20Law%20-%20Report%202019%20FINAL.pdf?updated=1593052831778.

Academic Papers

The technical report is generally the primary way to quickly distribute the findings from the Threatcasting Workshop so that you can empower individuals and organizations to change in order to prevent the future threats. However, that may only reach a percentage of potential audiences that have a role to play in disrupting, mitigating, and/or recovering from these threat futures.

An academic paper published in a journal that is appropriate to both the topic and intended audience is another possible delivery mechanism. The intent of the academic paper is not to cover all the topics in the technical report. Instead, it is to pick one or two findings from the Threatcasting work and expand for a slightly different audience.

An analyst(s) could choose to write a research article, technical article, or position article. The specifics for length and content are based on the standards for the academic journal the analyst(s) is targeting. Next are some general directional questions to help the analyst(s) decide where and what to publish.

General Directional Questions

- Who do you want to reach or empower about a specific threat future from the Threatcasting Findings?

- Are these individuals generally researchers or professors or students?

- Are these individuals within one specific academic domain or is it multi-disciplinary?

Select the journal before starting to write as many have very specific content formats, length, etc. Take the time to read some of the articles that have been recently accepted for publication so that you have a better grasp on how to write your article.

Annette Brown (2017) remarks (on why to publish in academic journals) that there are "three reasons: because it increases the credibility of the research, it increases the credibility of the practitioner, and it puts the knowledge in the permanent, searchable record. Spoiler alert: I think the third is the most important."

Resources

The following are examples of academic papers generated from Threatcasting activities.

- The Inside Enemy: Weaponisation of your logistical footprint—https://threatcasting. asu.edu/sites/default/files/2020-07/Johnson_Vanatta_The-Inside-Enemy.pdf.

- Contesting Key Terrain: Urban Conflict in Smart Cities of the Future —https:// cyberdefensereview.army.mil/CDR-Content/Articles/Article-View/Article/2420090/ contesting-key-terrain-urban-conflict-in-smart-cities-of-the-future/.

Executive Summary, Briefing, and Presentations

The Technical Report and the Academic Paper are generally longer in length and contain a high level of detail. Some audiences might not have the time, interest, or use for a long report. To quickly get across the Threatcasting Findings, an executive summary, briefing, or presentation could be best.

The goal of this form is to summarize the full report or paper with the highlights of the research, threat futures, findings, indicators, and actions. If the audience needs more information, they can and will reference the report or paper.

- Executive summaries are usually one to two pages if written.

- Briefings can be from 5- to 60-minute presentations.

- Presentations are typically no more than 9 or 10 slides.

All three of these areas are meant to provide people and organizations with a quick glimpse into the results and findings. If the reader or audience finds the executive summary, briefing, or presentation engaging and informational they can then reference the full report. "The summary should include the major details of your report, but it's important not to bore the reader with minutiae. Save the analysis, charts, numbers, and glowing reviews for the report itself. This is the time to grab your reader's attention and let the person know what it is you do and why he or she should read the rest of your …(report)" (Markowitz, 2020).

Briefings and presentations are different from a written summary because the analyst(s) will deliver them. There are many best-known methods for giving a briefing or presentation from business and industry.

Slides 1–3.

Describe the problem/opportunity. This is a research-based summary of the threats, strategy, challenges, and opportunities. This summary should be couched in the language that's used inside the organization or application areas.

Slide 4.

Summarize briefly who you are and your perspective as an analyst.

Slides 5–8.

Give an overview of your inputs, threat futures, findings, and indicators and actions.

Slide 9.

Ask for the next step. In most cases, this can be a delivery of the full report; a follow up, more in depth report out; or specific call out to the action items in the final report.[11]

Resources:

The following examples can help with drafting executive summaries, briefings, or presentations.

- Corporate/Business example: Mark Shonka and Dan Kosch (2002), co-authors of "Beyond Selling Value—A Proven Process to Avoid the Vendor Trap."

- Example: The Widening Attack Plane (sic). https://m.youtube.com/watch?v=HHD-DgkZOBFk.

- Executive Summary: (Page 2) COVID 19: Future Implications and Lessons Learned from COVID-19. https://threatcasting.asu.edu/sitcs/default/files/2021-03/Threatcasting%20Research%20Report.pdf.

7.1.3   ALTERNATIVE OUTPUTS TO DELIVER FINDINGS

The first three outputs are generally the most typical ways to capture the findings, however they are not the only way. They will not always capture the attention of senior leaders that don't have the time to devote to intensive reading nor room in their schedules for in-depth briefings. Therefore, there is a need to create a different format to express these rich and well-researched concepts and threats in a way that allows the intended audience to quickly grasp the human, ethical, political, or military impact and imagine the second and third order of effects that might follow.

Graphic Novels and Comic Books

A graphic novel or comic book is a visual representation of a story. The SFPs and short vignettes drafted for the report can serve as inspiration for a longer format visual version of a future threat. The goal of this visual medium is to engage the reader and allow them to imagine what these threat futures might feel like. Additionally, an introduction and afterword to the visual story can be used to highlight the broader findings upon which the graphic novel or comic were based.

---

[11] Adapted from: Geoffrey James. How to Present an Executive Summary. August 21, 2007. Moneywatch. https://www.cbsnews.com/news/how-to-present-an-executive-summary/.

The following (Figure 7.1) is an example of a graphic novel based on a Threatcasting Technical Report.

BUILDING A BETTER, STRONGER AND MORE SECURE FUTURE FOR OUR ARMED FORCES

Science Fiction Prototypes are science fiction stories based on future trends, technologies, economics and cultural change. The story you are about to read is based on threatcasting research from the Army Cyber Institute at West Point and Arizona State University. Our story does not shy away from a dystopian vision of tomorrow. Exploring these dark regions inspires us to build a better, stronger and more secure future for our Armed Forces.

Technology is evolving at a fast pace - the miniaturization of processing power combined with advancements in algorithms and sensor technology could radically change what our future looks like. It will be an era of accelerated human progress.

Therefore, we must have agile, adaptive organizations that continually encourage innovation while weighing the risk versus reward. We must embrace change in our thinking and actions in order to remain successful. This science fiction prototype is meant to spark these conversations.

Lt. Col. Natalie Vanatta
U.S. Army, Cyber

HERO

The year is 2028. Amid chemical attacks and global conflict, a rumored summit is about to convene in a secret location. Tensions are high. Rumors are rampant. A terror cell has infiltrated the summit and an attack is imminent.

Who can gather the crucial information needed and disarm the threat? It's time for an unlikely hero.

AFTERWARD

"A.I. [artificial intelligence] is today what aviation was 100 years ago. Like that other disruptive technology, we cannot ignore how A.I. will change our world."

Max Brooks
Best-selling author

In this story, our miniature hero was a vital member of the team - an autonomous robot searching for covert clues to prevent an attack.

What if in the future your teammates might not all be human? What if you worked beside a mechanical device, connected to a robust sensor network, that can process and analyze terabytes of data at machine speed real-time to develop an understanding of the environment that would take humans months or years?

While this might enable our victory, how do we also protect ourselves when our adversaries use it against us to exploit the initiative to create positions of relative advantage?

How does this scenario reshape your concerns about privacy? Should the definition of privacy evolve to accommodate the technology, or should we force technology to work within our current definition?

What might happen if our algorithms evolve past using statistical correlation to make decisions to being able to execute causational thinking? Then might we trust them to make decisions for humans without humans in the loop? Specifically from a law enforcement or judicial perspective, how would we integrate this technology into our legal system? Would a prosecutor be able to call Hero to the stand to cross-examine how they collected the evidence? Would the larger AI system that Hero feed its data into be able to articulate to the court how the investigation evolved over time and stayed within the scope of legal allowances?

While this graphic novel highlights both positive and adverse possibilities of future technology, our future does not have to be dismal. If we start thinking now, we can better prepare ourselves, our communities, and our organizations ..... how can you help? Your ideas could be the solution!

For more information or to share your ideas of how we need to adapt for the future, please contact us (threatcasting@usma.edu)

Figure 7.1: Hero (an example of a graphic novel by Johnson, 2018).

The graphic novella *HERO* explores a possible and potential future where artificial intelligence and robotics have been miniaturized to the point that a "bug" can fly into a secure area to conduct surveillance on potential terrorists. The visuals give the reader a sense of what this future could look like and how the "bug" might operate in the environment. More importantly, the graphic novella brings up serious privacy and rights concerns concerning the future of surveillance.

The aim of the graphic novella is to explore a small portion of the possible and potential threats but do it in such a way it illuminates the human, ethical, and legal implication of the future. If the reader is interested in the subject matter they can then be directed to the full report.

Resources:
The following references can help visualize what a graphic novel (based on Threatcasting) could look like.

- Invisible Force—https://threatcasting.asu.edu/invisibleforce.

- Dark Hammer—https://threatcasting.asu.edu/Dark_Hammer_Retrospective .

- Two Days After Tuesday—https://threatcasting.asu.edu/graphic-novella/cisco-two-days-after-tuesday.

- To Save Tomorrow—https://threatcasting.asu.edu/graphic-novella/future-work.

Interview/Podcasts

Like the short briefings mentioned previously, an interview or podcast can be another effective way to give an audience a high-level overview of the findings. Recently, podcasts, round tables, and discussions have become a popular medium to give audiences a high level of understanding of complex topics.

The goal of this medium is to pique the audience's interest and provide a link or connection so that they can read the longer version of the findings and other assorted works.

Resources
The following examples can help with crafting a podcast or interview about Threatcasting and/or its results.

- Mastercard: The Proactive Future of Cyber Security—https://m.youtube.com/watch?v=WhVLuWmaFCs.

- Cyber Security in 10 years—https://youtu.be/qu3Ut8eteKw.

- Cognitive Crucible—https://information-professionals.org/episode/cognitive-crucible-episode-14/.

7.2   POETRY

Editorial Notes

One note might be (she said)
to pull back somewhat [12]

Margaret Atwood

Why This Poem Matters

Atwood understands the power of storytelling and how important it is for people to experience the human impact. This is an interesting line to walk when pulling together your output, whether that be a technical report or a science fiction prototype. It is important to remember on a certain level to let the reader experience the human impact of the possible and potential futures that have been created.

Following Atwood's line of thinking, you can leave space to let the reader come to their own conclusions or more importantly to "own their conclusions."

---

[12] Exerpted from Atwood, M. (2020) *Dearly*. https://www.harpercollins.com/products/dearly-margaret-atwood or https://bit.ly/3okvdAO.

Project: "The Next Biological Public Health Crisis"

The Project section of this book will walk you through a shortened or abbreviated mock Threatcasting Workshop. The inputs, questions, and output have been simplified as a practice for each phase of the Method. Each Project section builds off the previous, just as the phases build off the previous phase in the Method.

PHASE 0—PREPARATION

For the purposes of the Project's mock Threatcasting Workshop, use the following Threatcasting Foundation.

> **Topic:** Future public health crises focused on virus-based pandemic (e.g., COVID-19 like) effects.
>
> **Research Question:** What are future threats from the next viral public health crisis?
>
> **Application Area:** What can government, specifically government-funded research labs, do to disrupt, mitigate, or recover from the effects?

PHASE 1—RESEARCH SYNTHESIS

Review Prompts

The following abbreviated prompts have been pre-selected for the Project mock Threatcasting Workshop. Read each of the following four prompts, pulling out key points that will be used to answer the Research Question.

Prompt 1: Social

The following points were pulled from multiple SME interviews with social scientists as well as public health professionals. The goal was to get a human understanding and perspective on the next public health crisis.

- Essential workers and under-served communities suffer high rates of Post Traumatic Stress Disorder (PTSD), suicide, domestic abuse, and addiction with minimal local and national support.

- Polarized political elections and party politics increase conflict between national, state, and local governments. The general public also becomes more polarized.

- The general public grows skeptical of government officials and traditional institutions.

  ○ However, the population is split 50/50 on whether they trust scientists, doctors, and public health officials.

  ○ There is overwhelming support for the armed services.

Prompt 2: Technical

The following points were gathered from SME interviews with Silicon Valley engineers, academic researchers, and journalists.

- Active and aligned government and private industry (e.g., tech and media companies) efforts reduce the spread of misinformation and disinformation about the virus over social media and internet sites to lessen the polarization effects on the general public.

- As the virus spreads and a wider swath of the population is hospitalized, the general workforce is diminished and infrastructure failures (supply chain, power, garbage, telco) increase tensions in the general public.

Prompt 3: Economic

The following points were pulled from SME interviews with economists, healthcare officials, and academic researchers.

- To avoid economic consequences state officials do not impose lock downs and stay at home orders to keep businesses open (restaurants, bars, salons, gyms, retail).

- With little local and national government actions, private corporations rally community outreach and charitable giving to social safety net and healthcare non-profits.

- Amid concerns of increasing corporate bankruptcies and workforce labor liability, businesses/corp individually close ranks, creating an inconsistent patchwork of responses to the national healthcare crisis.

- Virus variants and new vaccine side effects continue to stoke mistrust and destabilize the economy.

Prompt 4: Data with an Opinion

The following data points were pulled from a Public Health Expert SME interview.

- Uneven government attempts at virus mitigation prove to be ineffective. New cases and variants continue. Public death toll rises.

- Hospitals strain under cases with sporadic failures propped up by local and national uneven support.

- Nursing homes, jails, and schools see continued flare ups and clusters.

- Case reporting (e.g., infections, hospitalizations, deaths) is inconsistent. Local politicians contradict health experts. Local cities and counties continue to fight and push against state government mandates.

- Vaccine roll-out is uneven, generating mistrust in a growing number of the general population.

Research Synthesis Exercise

Using the four abbreviated prompts, conduct your own research synthesis exercise. The goal is to explore your notes about the four prompts as they relate specifically to the Threatcasting foundation.

Try to pull out as many data points and ideas from the prompts in response to the research synthesis questions as possible.

(NOTE: The following could be turned into a visual or made into an online form [similar to Figure 3.1]; whichever is easier for you to record the results of this exercise.)

- Data Point

  - Review the four prompts and your notes. As a participant, what data points stood out to you as important to the Research Question: What are future threats from the next viral public health crisis?

  - What data points would apply to the application areas: What can government, specifically government-funded research labs, do to disrupt, mitigate, or recover from the effects?

- Implications

  - For each data point you have written down, what are the implications to the Research Question: What are future threats from the next viral public health crisis?

  - To help refine the implications consider how they might apply to the application area: What can government, specifically government-funded research labs, do to disrupt, mitigate, or recover from the effects?

- Positive or Negative

  ○ For each implication, consider if it will have a positive or negative effect on the application area: What can government, specifically government-funded research labs, do to disrupt, mitigate, or recover from the effects?

  ○ Note: Your answer can be that the implication will be both positive and negative.

- What should we do about it?

  ○ Consider the research question: What are future threats from the next viral public health crisis? And the application area: What can government, specifically government funded research labs, do to disrupt, mitigate, or recover from the effects? Write down what actions the government, specifically government-funded research labs, could take to disrupt, mitigate, or recover from the next public health crisis.

PHASE 2—FUTURECASTING

For this mock Threatcasting project, you are asked to create a simplified EBM.

Pick Foundational Data Points

Directions:

1. From the list you create in Phase 1, select four data points. One should be from each of the four prompts.

   Note: You can pick them directly or you can roll dice, pick cards, or use a random number generator.

2. Write down the data points you selected. These will create the foundational structure for your EBM.

A Person in a Place Experiences a Threat

Imagine you are 10 years in the future and you encounter a person in a specific place experiencing a specific threat.

Directions:

1. Write down your person's name, age, and general characteristics.

2. Write down their occupation.

3. Write down three to five bullets with ideas on what their broader community might look like.

4. Write down what a typical day in their life might look like.

5. Write down where they might be when the threat event could occur.

6. Write down what the event is and briefly describe how your person experiences the event.

Experience Questions

Now that you have your person in a place experiencing the threat, use the experience questions below to expand the detail for the event.

Directions:

1. Write down when the person first encounters the threat. Date, time of day, day of week, month—any details that deepen the understanding of the threat.

2. Write down what they see. Describe it visually.

3. Write down if there are any smells or noises they attribute to the threat event.

4. Write down what the scene will feel like. Describe emotions.

5. Write down what they do not see or understand until later. Perhaps these are things subconsciously noticed.

Enabling Questions

The following enabling questions are meant to get you to look at the threat and the event from multiple perspectives.

Directions:

After thinking about what would need to happen to bring the threat about and how the threat actor or threat events from the virus might be enabled, think about the following.

1. Write down what existing barriers would need to be overcome to bring about the threat.

2. Write down what new business models and practices would need to be in place to enable the threat.

3. Write down what partnerships (or lack thereof) were in place to help enable the threat.

4. Write down what technologies (or lack thereof) were in place to help enable the threat.

Wrap Up

In this phase, you drew from the data developed in Phase 1. Additionally, you will have followed the methodology to populate an EBM with experience and enabling question answers based on the points from the RSW.

Review your Phase 2 answers, considering the initial prompts and the Threatcasting Foundation for this project.

- Would you add to, change, or expand upon your answers?

- Have you told the best story you can tell?

- Do you understand the story behind the story?

- Have you led with your imagination and pushed yourself to think from multiple angles?

- Would you adjust your language to be better understood by the application areas?

PHASE 3—BACKCASTING (TAD)

Based upon your Phase 2 EBM, answer the following questions.

Gates

- What are the actions that can be taken by the "who" in your application areas (the government, specifically government funded research labs) to disrupt, mitigate or recover from the threat? List at least one action that looks to disrupt the threat, one action that could mitigate the aftermath of the threat event, and at least one action that would assist the affected community in recovering from the threat event.

Flags

- What are the multiple iterative flags or indicators that could happen over the next decade that will indicate that the threat is manifesting? Be as specific as possible. You should be able to list at least three flags.

  ○ How will each flag build off the previous flag?

  ○ Can you add any additional flags that would enable the very first flag you have identified?

- Review your flags. Do your flags necessitate a branch in backcast?

  ○ Parties: Could a special-interest group lobby multiple government officials to pass a law and the law is passed in one state but not another?

  ○ Actions: What if a criminal cartel began to test and rehearse an upcoming crime but that crime was only possible in a handful of cities because of lax cyber security? Then those cities would branch from the rest of the world.

  ○ Location: Could a flag occur in one location (i.e., Country A) but not occur in Country B?

- Are there parallel future paths?

  ○ Are there any people or organizations that could take an action that would branch your backcast? Perhaps Company A (party) takes one (action) while Company B (party) takes another (action), setting them on different but parallel and competing backcasting paths.

Milestones

- Building off your gates and flags, now you can build the milestones.

- What can the government, specifically government-funded research laboratories, do in the next four years to prepare to disrupt, mitigate, and/or recover from this threat?

- Then what can the government, specifically government-funded research laboratories, do between years four to eight to disrupt, mitigate, and/or recover from this threat? Ensure that any actions you list in the timeframe of four to eight years

Once you have answered all the Phase 3 questions, review your answers.

- Can you add more detail to them?

- Now that you have completed the TAD Backcast, are there more flags that you had not originally thought about?

- Finally, do you need to adjust your milestone timeline, moving actions forward or backward in time?

PHASE 4—POST ANALYSIS

Given that the post analysis of the data is potentially the most challenging of skills for a new analyst, we have provided some mock EBM data for you to use along with the full EBM you created in Phase 3 of the project. This will also help to replicate the fact that you would, typically, have multiple EBMs to conduct the post analysis on. The three EBMs are entitled by their group name (Team Blaze, Team Inferno, and Team Scorch) and labeled as Figures P.1–P.3.

Review the project's Threatcasting Foundation before you begin. Create your post-analysis worksheet.

Conduct your first round of post analysis (summarize) on your EBM plus the three new ones with the guidance from Section 6.1.3. Record the results on your post-analysis worksheet.

- Review your summaries. Do they capture what the participants wrote?

- Would it be possible to only use the Summary for the rest of the Post-Analysis Phase and never return to the original EBM?

- Are there specific points in the EBM that make it unique? Did you capture that?

Conduct your second round of post analysis (meaning) on your EBM plus the three new ones with the guidance from Section 6.1.3. Record the results on your post-analysis worksheet.

- Review your thoughts on meaning. What interesting themes did you find when examining all four EBMs?

- Were you able to capture not just what the participants said but also what they "meant?"

- How does this section relate back to the Threatcasting Foundation?

- Does it give insight into the Research Question?

- Will it enable the application areas to apply to results?

Conduct your third round of post analysis (novelty) on your EBM plus the three new ones with the guidance from Section 6.1.3. Record the results on your post-analysis worksheet.

- What is new about this threat future?

- Is it a new idea for the Application Area?

- Does the novelty completely change how the Who in the application area will think about the research question and threat?

- Will the results enable action for the Application Areas?

- Is your novelty an outlier?

Example EBM 1—Team Blaze

Project: Example EBM 1	
TEAM Blaze	
SCENARIO TITLE:	Everyone step forward...not so fast, United States
DATE OF FUTURE EVENT (year):	2031

Destabilization Scenario	
Where will it occur?	Texas, US (which is planning its secession in order to escape the US barricade)
Who will it affect? Describe your person, in a place.	Truck driver who lives in El Paso, makes frequent border crossings for interstate and international commerce, and parents still live in Ciudad Juarez, MX.
When do you expect it to happen?	2031

Experience Question	
Describe the "event". How will it unfold?	The emergence of a novel Ebola-Marburg variant is the first pandemic since COVID-19. The world reacts by shutting down borders and amassing its collective militaries around America's borders -- including land, sea, air, and space -- activating a secret global network to protect the world from the United States and a repeat of the national leadership failures of 2020. They didn't want to activate the network, but were forced to when the American President was overthrown by the conspiracy theory party that is certain that she is a lizard person.

Enabling Question	
What led up to it? What were the factors that brought it about?	We will have instant tests that are able to detect instances of coronavirus, but nothing for Ebola. Because COVID-19 only lasted for 3 years in the US, a full public health infrastructure was not built. The President lost the Senate and opposition Senators zeroed out the public health funding for the United States to enable funding national conversion therapy camps to reprogram gay youth and train them for filibustering. As the virus spreads, healthcare workers are triggered with memories of 2020 and start dropping like flies. The healthcare system is overwhelmed in a matter of weeks.

Backcasting		
Gates		
	1	Efforts to grow PPE manufacturing infrastructure in United States failed due to lack of federal funding and waning interest by the private sector in manufacturing. The tech industry continues to grow political power and pushes commitment to globalization.
	2	International tourism reaches new heights with pent up demand for travel after COVID-19.
	3	WHO collapses to regional health organizations. (Turkey and Israel aren't invited to sit at anyone else's lunch table. They form the falafel alliance.)
	4	Burnout and early retirement post-COVID decimates the US healthcare and pubic health workforces.
	5	Rapid screening tests that detect novel viruses/variants.
	6	Broad spectrum antivirals, widely available and accessible.
	7	Low-cost, widely available viral sample gene sequencing. (surveillance)
	8	Thermal sensing technologies are common in the built environment.
	9	Increased funding and collaboration to implement safer workering/living conditions for communities near animals (preventing zoonotic diseases)

Flags

1	Emergence of Regional Health Organizations (RHOs)
2	Countries prioritize funding RHOs and WHO collapses.
3	Official formation of a new 3rd political party with candidates on 2024 ballots.
4	Down-ballot offices held by new 3rd political party members.
5	Excess deaths in the Amazon
6	Simultaneous collapse of pan South American governments
7	Ubiquitous health screening in public buildings and private businesses
8	Development of (more) underground anti-government networks.

Milestones (4 years)

1	Case of novel Ebolaburg virus detected in the Amazon. Sentinel surveillance system didn't pick it up until 4 tourists were infected and had already traveled to their home countries.
2	Unchecked energy crisis precludes adequate deep freezer storage for mRNA vaccines.
3	In Texas, new 3rd political party and Secession party members are elected to the state assembly and form a coalition government. (Decrease social media's tolerance for conspiracy theory accounts and mis/disinformation)
4	Viral sensing technologies replace thermal sensing in the built environment.
5	States buy their own PPE supplies from foreign manufacturers. Border states have high PPE costs and relatively low supplies. Border state governors justify low PPE expenditures by demonstrated herd immunity.
6	Viral detection construction materials.
7	More resilient mRNA vaccines that do not require freezing.
8	Effective detection and disruption of conspiracy theory networks.
9	Non-pharmaceutical in-vivo antiviral technologies.

Milestones (8 years)

1	Ubiquitous broad-spectrum antiviral screening (noting that this is unlikely to go over well and will likely lead to massive eruption of conspiracy theories and anti-government networks).
2	Higher federal and state requirements for energy production from sustainable sources.
3	Restoration of State Dept funding with clear focus on diplomacy as the primary means of international engagement.
4	US and international focus on widespread sustainable development in the Global South
5	Establishment of volunteer U.S. White-Hat Hacker Force. (Their uniform will be t-shirt, hoodie, baggy jeans, and vans or Chuck Taylors). They are paid in Bitcoin and pizza delivery with vegan options.
6	Social scientists liberally located throughout national lab, IC, defense sectors.
7	Highly efficient and affordable sustainable energy production.
8	Adequate funding for cybersecurity professionals in fed government.
9	Adequate funding for emerging infection surveillance federally.
10	Increased funding for healthcare and public health workforce development and retention.

Figure P.1: EBM created by Team Blaze. (From Brown (2021), used with permission.)

Project: Example EBM 2	
TEAM Scorch	
SCENARIO TITLE:	Come Die In the Desert!
DATE OF FUTURE EVENT (year):	2031
Destabilization Scenario	
Where will it occur?	Arizona
Who will it affect? Describe your person, in a place.	Social organizier/activist (Lauren Kuby - who rage quit her Tempe city council job after the Macayo's debacle of 2027) is trying to sponsor a ballot initiative, Prop 12635, that would finally force Arizona to allow communities to enact and enforce local universal public health guidelines. After ten years of failed and lacking public health policy at the state level, Prop 12635 was written by a coalition of environmental activists inspired by their own inabilities to live is safe, healthy communities constantly attacked by poor air quality, annual innoculations for newly emerging covid variants, and lack of action by contentious state and county politics. The cornerstone of the proposition is the establishment of "Safe and Healthy Communities" (SHC's) that meet the "Kuby Standard" for geographically protected communities to share in protection from infectious disease and environment assaults from people who "frankly don't give a shit about their fellow humans". SHC's as envisioned in Prop 12635 will be recognized by municipal health departments (which the prop establishes and redirects 90% of federal funding to municipal gov'ts on a per capita basis) as long as they can demonstrate enforcement by municipal police. Annual reporting is required. According to Kuby, "This initiative will finally allow people who want to live in a safe, healthy community to exercise that right without interference from bad-actors and immigrants who are attracted to Arizona's lawlessness". The proposition has already gathered enough signatures to appear on the 2031 ballot despite staunch opposition from the AZ Commerce Authority and the "Arizona is Open" PAC. Opponents cite that the initiative is illegal, will cost taxpayers billions of dollars in lost revenue from economic expansion, and will only hurt small businesses who rely on the state for essential tax relief.
When do you expect it to happen?	Early 2030s
Experience Question	
Describe the "event". How will it unfold?	Several new community/advocacy groups spring up in response to the Arizona government's extremely lax enforcement of public health measures. The past decade has made it clear: despite rolling stay-at-home orders and public health regulations in other states, AZ is a haven for refugees from oppresive states who are looking for normal lives free from rules and restrictions. The advocacy groups are mobilizing toward a campaign (Prop 12635) that will get a statewide ballot initiative to establish public health departments in every city, establish a statewide mask mandate, and have law enforcement consequences/enforcement for noncompliance. This would empower cities to have their own health departments and enforce stricter rules than what the state enforces. Cities will adopt SHC's with voluntary but 100% compliance, with limited zones of entry/re-entry. Since corona-viruses and other infectious diseases have been removed from insurance schedules in 2022, the federal government provided tax credits to private healthcare companies to build medical campuses and hospitals to handle the increased burden of respiratory and other environmental disease caused by poor environmental conditions. SHC's have negotiated discounted rates to recognize their contribution to reducing risk within their responsible communities.
	The bill will promote the establishment of virus-safe SHC zones (LEED-like standards about ventilation, design, public health rules dubbed "Kuby Standards"). Many people try to tie these virus-safe to other environmentally-friendly movements claiming that they don't go far enough, but there are points of conflict between them (ie. cars vs public transport, resusable vs disposable utensils, cleaning standards for waymo, segregated grocery stores and dining areas). Many smaller cities and neighborhoods claim that the only solution is Bubble Cities with complete perimeters and dedicated services.

Enabling Question	
What led up to it? What were the factors that brought it about?	Arizona never imposed restrictions to get COVID-19 under control - in lieu of mask mandates and public health measures, health system has been expanded and privatized. Communities without access to healthcare start to set up treatment in homes and non-medical facilities. There are now 9 known strains of coronaviruses in circulation. Vaccines are annual but strains still emerge that get around them and people have varying degrees of adherence. Advertising from pharma companys is now direct-to-consumer, creating a patchwork of vaccinations for those who can afford it. AZ continues to see massive population growth, despite increasingly hostile climate-change driven conditions in Phoenix (hot weather, water stress, air pollution). Arizona attracts people who value personal freedom due to the lack of restrictions.

Backcasting	

Gates

1	Virus Screening Tools - Could play a role in having Arizona """safely""" open. Real-time tests are common and widespread. As SHC's are being planned, local businesses are using rapid screening tools and mobile apps to identify the most recent test result and other biometrics as condition for entry.
2	Anti-viral surface treatment - Could be a point of investment for businesses and buildings (hygiene theater). One of the regulatory requirements for Prop 16235 is the maintenance of approved surface treatment requirements for businesses, cars, public transportation, and homes (yes, homes).
3	Aerosol/fluid dynamics - Implications for upgrading indoor spaces wrt filtration and ventilation. Establishment of virus-safe zones based on upgraded design and public health elements (to Kuby Standard level). New mask technologies? Several local architectural firms have begun developing whole-home sterilization systems that require little energy and meet the Kuby Standards. Exercise and fitness studies have personal oxygenated masking stations.
4	Medical countermeasures - Better treatments for virus-enhanced diseases exist (but strains of COVID-19 continue to claim between 25,000-100,000 lives/year in the US). Direct to consumer marketing has increased vaccine uptake considerably, improving individual decisions! A huge win for public health!
5	Models predicting or monitoring transmission - Newly minted municipal health departments (that can afford it) make use of sophisticated modeling software to track the transmission of disease, informing better decisions! Mobile applications enable citizens to keep all of their data safely anonymous while enabling complete predictive situational awareness at the building level to protect SHCs. Syndromic surveillance enables SHCs to raise protection levels if there are eminent threats from their surrounding, non-participating residents.
6	PPE Improvements/Sterilization - Likely confined to healthcare space (could play a role in beefing up healthcare system), unless sterizing N95s becomes a consumer product
7	a)How do these programs work for people who can't or won't participate in them (whether due to lack of economic access or personal beliefs). b) Will this place undue burden on small businesses? c) Gentrification d) Public health workforce - where are we going to get them in every municipality. What about rural areas - is there a size cutoff for who has to establish a new health dept? e) What will this mean for municipalities that don't want to impose stricter rules. Will they still have to establish a municipal health department too? Argument that this is govenrment waste?

Flags

1	This is a slow boil scenario. General malaise from the public about inadequate health care and health protections. Increases in privatization of AZ communities (a la homeowners associations).
2	The first confirming sign that covid vaccines will not deliver 100% permanent protection against covid, a different emerging pathogen, or quickly waning immunity and re-infection that necessitates ongoing lifestyle amenities to protect from disease outside of healthcare settings.
3	Rural vs. urban divide becomes ridiculous as most rural areas don't understand the attempts to block out nature.
4	Lack of interest in safe, healthy communities is rampant, leading to little change outside of local advocacy communities.
5	Environmental countermeasures (hygiene theater) to airborne viruses become widely accessible as consumer products. These countermeasures become commonplace within businesses that can afford to upgrade.
6	Specific upgrades to ventilation and other aspects of design emerge as effective airborne virus countermeasures. Virus-safe design, upgrades, and amenities become desireable (if more expensive) real estate features
7	Community organization and the formation of advocacy groups around their rights, despite a lack of consensus. (Public health organizers vs people opposing them for various reasons)
8	Privatization, including the fundraising for political action groups, taking positions seeking to privatize and fragment healthcare, environmental safety (avoidance) and reduce risk.
9	A bunch of Karen's get really upset and start meeting in person. (And also start calling themselves The Resistance)
10	Vigilante groups and organizations taking direct action against laws and regulations due to a lack of care for enforcement.

Milestones (4 years)

1	Virus Screening - Could be used in addition or as an alternative to architectural virus countermeasures
2	Surface Treatments - Could play a role in reopening if deployed as a consumer product or were able to be picked up by businesses
3	Aerosol/Fluid Dynamics - better understanding of standoff zones, infection probabilities, and environmentally engineered risk.
4	Medical Countermeasures - Better anti-viral treatments make the prospect of getting COVID or other airborne virus less likely to have severe consequences
5	Modelling transmission - Will models have any real-world impact if an outbreak doesn't spur any additional action or safety measures. Could outbreak info be made available to citizens so that they can make their own decisions?
6	PPE improvements - Better masks, perhaps some types of PPE become standard for staff outside of hospitals
7	Protection for immunocompromised people + Balancing of virus infected patient care with other hospital care
8	Continuation of COVID conspiracies that carried over from 2020?
9	What happens to people who are disabled by viral infection diseases like COVID-19?
10	Vaccine distribution infrastucture + promotion?

Flags	
1	This is a slow boil scenario. General malaise from the public about inadequate health care and health protections. Increases in privitazation of AZ communities (a la homeowners associations).
2	The first confirming sign that covid vaccines will not deliver 100% permanent protection against covid, a different emerging pathogen, or quickly waning immunity and re-infection that necessitates ongoing lifestyle amenities to protect from disease outside of healthcare settings.
3	Rural vs. urban divide becomes ridiculous as most rural areas don't understand the attempts to block out nature.
4	Lack of interest in safe, healthy communities is rampant, leading to little change outside of local advocacy communities.
5	Environmental countermeasures (hygiene theater) to airborne viruses become widely accessible as consumer products. These countermeasures become commonplace within businesses that can afford to upgrade.
6	Specific upgrades to ventilation and other aspects of design emerge as effective airborne virus countermeasures. Virus-safe design, upgrades, and amenities become desireable (if more expensive) real estate features
7	Community organization and the formation of advocacy groups around their rights, despite a lack of consensus. (Public health organizers vs people opposing them for various reasons)
8	Privatization, including the fundraising for political action groups, taking positions seeking to privatize and fragment healthcare, environmental safety (avoidance) and reduce risk.
9	A bunch of Karen's get really upset and start meeting in person. (And also start calling themselves The Resistance)
10	Vigilante groups and organizations taking direct action against laws and regulations due to a lack of care for enforcement.

Milestones (4 years)	
1	Virus Screening - Could be used in addition or as an alternative to architectural virus countermeasures
2	Surface Treatments - Could play a role in reopening if deployed as a consumer product or were able to be picked up by businesses
3	Aerosol/Fluid Dynamics - better understanding of standoff zones, infection probabilities, and environmentally engineered risk.
4	Medical Countermeasures - Better anti-viral treatments make the prospect of getting COVID or other airborne virus less likely to have severe consequences
5	Modelling transmission - Will models have any real-world impact if an outbreak doesn't spur any additional action or safety measures. Could outbreak info be made available to citizens so that they can make their own decisions?
6	PPE improvements - Better masks, perhaps some types of PPE become standard for staff outside of hospitals
7	Protection for immunocompromised people + Balancing of virus infected patient care with other hospital care
8	Continuation of COVID conspiracies that carried over from 2020?
9	What happens to people who are disabled by viral infection diseases like COVID-19?
10	Vaccine distribution infrastucture + promotion?

Milestones (8 years)	
1	Advanced screening technologies that can identify infected people being used for real time entry/access into smaller and smaller spaces: neigborhoods, businesses, communities.
2	Healthcare treatments for individuals reach mutliple tiers and locations with highly divergent levels of care - both public, private, municipal, and community. Including the introduction of communities built around healthcare.
3	Consumer products allowing for indoor sterilization and safety become accessible
4	Architectural/ventilation standards for safe indoor spaces become more established
5	The Kuby Certification for Safe, Healthy communities - includes air quality and infectious disease standards (cleaning, surveillance, testing, and allowed persons) /to deal with continuing emerging viral strains and climate problems at the same time.
6	Community based modeling: actionable information used to make information about local decisions, such as how many people to allow to move into communities, when to open/close, and when to enforce constantly changing rules and standards to protect safety.
7	Privitization and lack of state govt action to address public health leads to need for communities to take things into their own hands in the first place
8	Prevention/mitigation of climate change issues (heat, water, pollution)
9	Focus on Phoenix Metro/Tuscon vs other parts of the state. How to establish healthcare instrastructure/workforce outside of metro areas. Similar question for poor/marginalized/tribal communities
10	What about Arizona's large elderly population (Sun City vs smaller communities within municipalities)? Barriers to access for them despite being highest risk?
11	Federal response options (what will that look like under 2021 President and whoever succeeds him?)

Figure P.2: EBM created by Team Scorch. (From Brown (2021), used with permission.)

Example EBM 3—Team Inferno

Project: Example EBM 3	
TEAM Inferno	
SCENARIO TITLE:	Your Brave New World featuring Pandemic Polly
DATE OF FUTURE EVENT (year):	Culmination in 2031
Destabilization Scenario	
Where will it occur?	New York City
Who will it affect? Describe your person, in a place.	Dr. Spock - Chief Medical Officer for Respritory Disease at Columbia University Irving Medical Center ,
When do you expect it to happen?	2021 - 2031 and beyond

Experience Question

Describe the "event". How will it unfold?

10 years later, COVID-19, now just called the "COVID flu," has become an endemic seasonal illness along with seasonal flu. In the intervening years, there have been many emerging strains of COVID and flu and adjustments to vaccines, but less focus on NPIs. A new betacoronavirus emerges and begins to spread rapidly, appearing at first to be the seasonal wave of the COVID flu. Initially, testing fails to distinguish between the new virus and circulating flu strains, until routine sequencing for new variants identifies the new virus as not COVID flu, causing mass panic.

In an effort to combat misinformation and fear about pandemics, and to infuse factual data into social circles, the CDC created age appropriate curriculum for K-12 students. A few of the characters in the introductory series for the youngest students are Pandemic Polly, Eli the Epidemiologist, Nurse Nancy, Veronica Virus, and Dr. Duke. Together they explain the basics about pandemics, viruses and varients, and provide the students with information on how they can be heros for healthcare workers and EMS providers.

Healthcare capacity has been reduced dramatically by PTSD and the long-term impact on in-person healthcare systems. Because of this, more systme shave to telehealth for day-to-day care and screening. Focus on telehealth makes it harder to distinguish between overlapping respiratory pandemics. Reduction of point-of-care testing, because people have moved to telehealth reduces speed of detection.

Public health systems have moved to tech-assisted models due to shortages in highly-trained after many left the field due to COVID-19. Medical and public health systems have entire EHR/EMRs and communicate digitally. Because systems have moved to telehealth, there are fewer centralized resources in place in offices and large healthcare systems. Cybersecurity is reasonable, but not appropriate for critical infrastructure. As healthcare has become increasingly digital, it has also increasingly connected to the Internet of Things, creating new vulnerabilities faster than they can be addressed.

The new virus is more transmissable to PPE (actually airborne), requiring improved PPE. While PPE systems have improved, they are still not capable of mass-producing updated PPE at speed to fulfill need. Creation and testing of improved PPE for the public is slow and stockpiles of old PPE are inadequate, allowing the market to be flooded with old and inneffective PPE before the new versions are available. Confusion about effectiveness and PPE types ensues.

Modeling the new virus and its spread is complicated by failures to differentiate between older COVID-19 strains and the new virus. Modelers attempt to work with incomplete data leading to incorrect predictions and destroying trust. Rapid sequencing has remained underfunded, meaning we do not have capacity to rapirdly track spread until people begin dying in new areas. Creation of rapid tests is complicated by existing ongoing seasonal disease.

Governments are slow to recognize and admit a new virus is spreading, due to memories around COVID-19. Populations have not regained trust in systems after the COVID-19 failures. An unregulated internet allows misinformation and conspiracy to spread around disease origins and severity (again). As tensions around the new virus, countries accuse each other of releasing it, leading to escalating cyber-warfare. This culminates in a Chinese cyber-attack on US systems that disables telehealth and public health systems, the power grid, and banking infrstructure. Attempts to rapidly repair systems fail, leading to spotty outages and unreliable power, internet, banking, and health systems.

Civil unrest related to lack of trust and failing systems increases viral spread and overwhelms healthcare and law enforcement.

Enabling Question

What led up to it? What were the factors that brought it about?

PTSD among healthcare and public health workers reduces capacity, technological solutions and movement to smart city and IoT infrastructure, lack of improvements in cyber-security, fatigue from dealing with ongoing seasonal COVID and flu outbreaks

Backcasting

Gates

1	Develop and maintain national distributed rapid sequencing capability and computational ability to compare sequences and identify new variants or viruses in real time
2	Improve cyber threat analysis and identify vulnerable national systems to focus on healthcare and public health
3	Improve modeling of viruses to understand seasonal dynamics with vaccines
4	Improve rapid PPE tesing and development, both professional and at-home, and communication of those results directly to the public. This should include attempts to limit fake PPE on the market and how the public and institutions can identify it independently.
5	Widely disseminate any novel computational tools developed; especially in regards to putative variants of infectious disease agents in active circulation.
6	Mental health interventions now and long-term for medical and PH personnel
7	Strategy for distributed sequencing efforts and defined performance of work sites
8	PPE Import Monitoring to prevent importation of fake PPE
9	CENTRALIZED database of PPE and corresponding protection level provided - from homemade to NIOSH Cert
10	Create long-term education campaigns about diseases and how to prepare for outbreaks, beginning at a young age and continuing through school (Zombie preparedness, Pandemic Polly)
11	Perform comprehensive after-action analysis on successes and failures of all systems and their integrations to prevent repeat mistake and inform future actions. Constant course correction based on analysis of past actions would enable long-term improvements

Flags

1	Reductions in healthcare and public health workforce due to PTSD and burnout
2	Moves to augment reduced workforce with telehealth and IoT devices (increased dependence on technology to make up workforce gaps)
3	COVID becomes an endemic seasonal disease with breakthrough variants that can evade the vaccine
4	Public response to viral breakthrough outbreaks becomes dulled by familitarity and rapid release of boosters for new variants
5	Increasing fragility of technology around telehealth and EMRs, including vulnerability to cyber attacks, reduces public faith in infrastrure ability to maintain privacy
6	Failure to address vulnerabilties and increasing creation of new vulnerabilitites leads to increased frequency of test cyberattacks by state actors
7	Modeling can track seasonal variants and begins to see changes in the virus patterns but can't address them or fails to communicate them appropriately
8	Public mistrust in institutions leads to doubt of PPE protection and use, failure to believe model results, and conflicting responses to information

Milestones (4 years)	
	1 Develop comfortable, effective, and rapidly scale-able N95-level PPE that could be provided to the public in an emergency and can be manufactured in multiple adapted facilities
	2 Enable ongoing automated cyberthreat testing and analysis for critical healthcare and public health systems and their connections to IoT devices. Harden these connections over time and perform constant testing of all new systems
	3 Ensure ongoing monitoring of breakthrough outbreaks and modeling of expected levels of disease for new viruses and flu combined to ensure the ability to detect deviations from expected diseases patterns and detect potential new respiratory viruses despite increased background noise
	4 Maintain functional investment in current testing infrastructure. Valuable investments in long-term infectious disease surveillance measures is crucial. Improvements to syndromic and genetic surveillance could detect emerging viruses early.
	5 Apply epidemiological modeling techniques to the spread misinformation and conspiracy (around outbreaks in particular) to enable rapid response to new patterns and interventions in social narratives before they increase transmission or conflict.
	6 Improve reslience to cyberattacks for power infrastructure. Improve ability to isolate sections of power grids and create power zones during moments of signficant instablity
	7 Combatting online misinformation and finding a way to increase the availability of non-politically motivatied scientific information. Address algorithms that drive people to increasingly fringe information in science, politics, and social issues.
	8 Perform comprehensive after-action analysis on successes and failures of all systems and their integrations to prevent repeat mistake and inform future actions. Constant course correction based on analysis of past actions would enable long-term improvements
Milestones (8 years)	
	1 Rapid provision of functional PPE and scale up of new PPE types to address the new virus. Open communication about PPE success and failure to the public to improve trust and use. Innovation in PPE to make it reusable, cheap, and universally available.
	2 Enhance predictive modeling to capture frequent changes in variants and quick identification of new viruses. Develop communication messaging to ensure highest quality of public dissemination.
	3 Constant improvement in cybersecurity around health to prevent potential future vulnerablitites and attacks by state and non-state actors
	4 Reinforcement of power systems and grids to prevent potential attacks, including establishment of backup capacities at crucial points to prevent loss of banking, medical, and other CISA infrastructure systems
	5 Better research into social dynamics in the face of outbreaks and emergencies would allow more effective responses to misinformation and social unrest caused by power outages and loss of system integrity
	6 Perform comprehensive after-action analysis on successes and failures of all systems and their integrations to prevent repeat mistake and inform future actions. Constant course correction based on analysis of past actions would enable long-term improvements

Figure P.3: EBM created by Team Inferno. (From Brown (2021), used with permission.)

It is also typical that there might be more than one analyst reviewing all the EBMs. Therefore, we have provided additional analysts' post-analysis work for this project.

Analyst A and B both used Team Blaze, Scorch, and Inferno to conduct their post-analysis activities. Along with their post analysis (Figures P.4–P.7), we have also provided their commentary to help explain their thinking at each of the three rounds of Phase 4. These two analysts (referred to as analyst A and B) apply different approaches to the set of EBMs (Figure P.1–P.3) which were adapted from a project investigating the future of public health following an incident such as the COVID-19 pandemic of 2020–2021. The method of how this analysis was conducted can be found in Jason Brown's doctoral dissertation (2021).

Additional Analyst(s) Techniques

The additional analyst commentary also demonstrates the flexibility of the post-analysis process to adapt to different analytical styles. Analyst A tends to bounce back and forth between rounds and holds a looser interpretation of the boundaries between each. Analyst B favors a stricter boundary between rounds and attempts to ensure each round is complete before moving on to the next. Finding your own approach is a matter of experimentation and practice.

Round 1—Additional Analyst

In this exercise, both Analyst A and B decided to use the short form summary and rewrote each scenario in a short paragraph. See Figure P.4 for how Analyst B wrote their summaries for the three provided EBMs.

	A	B
1	**Group**	**Round 1 "Summary"**
2	Team Blaze	An isolated US forced to fend for itself when global blockade is thrown up around borders to stop the spread of a novel Marburg-Ebola variant emerges. Instant testing for the last pandemic provers ineffective for the next one - Public infrastructure is weak and has not been reconsidered in aftermath of COVID-19 and national funding is inadequate to deal, having been redirected to irrelevant programs.
3	Team Scorch	Failure of national and state public health policy guidelines or reform post-Covid-19 have led to communities seeking (via legislation) the ability to self-police and enforce public health requirements. With national and state authorities choosing not to engage or actively enforce meaningful health guidelines, concerned citizens must mobilize to do it themselves. In this scenario, they are doing it in Arizona, which happens to also be where where peope seeking freedom from regulation have been migrating.
4	Team Inferno	In 2031 - exhaustion and fatigue related to ongoing endemic illnesses; attempts to educate population with factual info related to pandemics become k-12 educational curriculae; Underfunded national/state healthcare infrastructures drives a shift to distributed and increasingly vulnerable telemedicine that makes idenitifcation of new strains difficult, Covid PPE nbot adequate for new variant, Active misinfo by politicians and bad actors creates confusion; rapidly evolving situation means lack of clear info - a vaccum filled by misinfo and distruast; Critical shortage of medical workers following Covid-19 PTSD thins ranks; govt looking to shift blame uses foreign actors, culminating in a crippling infrastructure attack by China - . Significant civil unrest outbreaks due to lack of trust and failed system response.

Figure P.4: Analyst B summarizes each model in a short paragraph that highlights the events and problems being investigated in each scenario. (From Brown (2021), used with permission.)

Round 2—Additional Analysts:

Both Analyst A and Analyst B used different techniques to record their round 2 results based on how they interpreted the data.

In Vivo coding

For example, Team Scorch (Figure P.2) imagined the idea of "*Safe and Healthy Communities*" (SHCs) that encompasses a larger concept where municipal governments that better know their communities should be empowered to create health and safety guidelines without federal or state level interference.

In another example of in vivo coding, you might use the phrase "*Reduction in point of care testing*" found in Team Inferno's model (Figure P.3) to describe the combination of a shift toward telemedicine and the lack of collaboration between telemedicine companies, which ultimately reduces the speed of detection for new viruses.

Gerunds

For example, Team Blaze (Figure P.1) identified several flags that described governments creating and funding regional health organizations as the World Health Organization (WHO) collapsed. "*Regionalizing health services*" could be a way to describe these flags.

Similarly, Team Scorch (Figure P.2) described an even more granular reduction in national level oversight of healthcare and emergency response. States (Arizona is the example used in Team Scorch) would take action at the municipal level to safeguard their population and would direct 90% of federal funding to municipalities that could demonstrate a certain level of safeguarding. Analyst A called this "*Localizing health services*."

In the Team Inferno model (Figure P.3), the scenario was imagining the social and professional exhaustion caused by an endemic illness. Without adequate policies to relieve healthcare workers of the fatigue of working under strict conditions, the healthcare system was beginning to collapse. Several examples of this pending collapse centered around personal protective equipment (PPE), educational confusion, and leadership failures.

Text from Team Inferno describing the PPE problems: "While PPE systems have improved, they are still not capable of mass-producing updated PPE at speed to fulfill need. Creation and testing of improved PPE for the public is slow and stockpiles of old PPE are inadequate, allowing the market to be flooded with old and ineffective PPE before the new versions are available. Confusion about effectiveness and PPE types ensues."

Analyst B focused on the technology aspect of preparedness and described the meaning of this idea as "*PPE inadequate for new variant*."

Figures P.5 and P.6 show different ways to think about finding Meaning: one from Analyst A and one from Analyst B.

	A	B
1	**Group**	**Round 2 "Meaning"**
2	Team Blaze	religion over public health
3		fallout of health care system
4		regionalization for necessary / health services
5		state segregation in favor of regional control
6		resurgence of far right conserviative values - GILEAD
7	Team Scorch	localization of health services
8		state segregation in favor of regional control
9		dissolution of united states into fiftoms of sanctuary or oppression based on public health mandate strictness
10		privatized solutinos
11		sanctuary cities for viral diseases
12		privatized tech emmergence of protective immediacy
13		divide of haves / have nots as well urban / rural in terms of adoption
14		smart home environment upgrades, mean home is a bunker a safety net. Instead of being trapped there, people will self-seclude for their own protection.
15		monitoring network to ensure an pandemic doesn't break
16		continuation of covid consipiracies
17		belief of publicly funded health care has gone down
18	Team Inferno	routine targeted sequencing
19		proactive early education for public health
20		assumes that telehealth is still just consultative
21		automation / tech assisted mechanisms are more widely accepted
22		fewer centralized medical data resources
23		reduction in health care career involvment necessitating supportive measures

Figure P.5: Round 2 "Finding Meaning" from Analyst A. Note the bullet points and short statements that illustrate how the Analyst decomposes the model into individual building blocks. (From Brown (2021), used with permission.)

	A	B
1	**Group**	**Round 2 "Meaning"**
2	Team Blaze	creating tools to address the last pandemic can blind us to the needs of the next; The unique geopolitical role of the US requires pandemic transparency and communications not just domestically but globally - going it alone risks global isolation; honestly confronting national, state and local pandemic
3	Team Scorch	potential fragmentation/civil unrest/secessionist/polarization; in-state "native" vs "newcomer" schism; Tribal issues and elderly-at risk population issues; conflict between a population that believes in collective responsibility (strong as weakest link) vs complete individual freedom
4	Team Inferno	shift to remote health compromises speed to identify new variants; *Healthcare workers avails vs PTSD post Covid; trying to reapply old lessons doesn't work - PPE inadequate for new variant - creating tools to address the last pandemic can blind us to the needs of the next*; during covid-19 when political infrastructure and health care systems failed us, leadership redirected blame to China (fair and unfair); Chinas response was soft power misinfo but this time, that wasnt enough this go around

Figure P.6: "Finding Meaning" from Analyst B. Note they chose slightly longer descriptions that began synthesizing a few ideas together and interpreting trends. (From Brown (2021), used with permission.)

Round 3—Additional Analysts

Analyst A combined the ideas (codes) of "localization of health services," "privatized solutions," "smart home environment as a bunker," and "privatized tech emergence of protective immediacy" into the novel idea of "*Prevention has become a business.*" This includes all the implications of using a business model with returns on investment and profit margins as a way to manage public health infrastructure and prevention. Although this model is familiar to insurance industries that mitigate financial loss after an event, it could be novel to municipalities and new businesses trying to capitalize on new localized public health infrastructure processes.

Another example, Analyst A discovered that concepts between the three models suggested the idea of "*Medical sanctuary (bubble) cities based on health, religion and data beliefs.*" While the concept of a sanctuary city is not new (consider sanctuary cities for refugees), the idea of providing sanctuary based on data beliefs, especially health data tied to religious beliefs, is a novel adaptation and might warrant significant attention in order to avoid, mitigate, or recover from the threat future pandemics hold. Analyst B (Figure P.7) reinforces the idea of sanctuary cities but calls them "*freedom destination states*" that are places for people to relocate to reclaim perceived losses in freedom from federal oversight.

Another example of novelty builds on the history of military recruitment. Analyst A recognized that the fatigue and political burnout of the healthcare industry might be mitigated by a "*much more aggressive recruiting stance and might provide similar benefits to military service in order to incentivize."* This describes one way the entire healthcare (and health prevention) system might adapt over decades to incentivize more participants and give them certain social status benefits that the military currently enjoys. "Thank you for your service" may no longer only refer to those fighting wars on foreign soil!

	A	B
1	**Group**	**Round 3 "Novelty"**
2	Team Blaze	National/state/city "truth and reconciliation" required to assess breakdowns in identification/care/response to best prepare for next. Will need to address clemency and leniency.
3	Team Scorch	"Freedom" destination states - places where people relocate for perceived freedom, away from states prioritizing collective responsibility - to the extent these become identitarian, expect conflict/secession, perhaps along tribal lands type models. Will US fight to keep authority?
4	Team Inferno	Geopolitical adversaries actively introducing pandemics for political, economic and military gain means we need a system to fingerprint viruses nationally and locally; might we create a human "canary in the coalmine" CCC group paid to submit to regular testing for pandemic?

Figure P.7: Analyst B describes the novel findings from their analysis. Notice that there are deeper implications listed in the form of questions that the data does not address, but the analyst recognizes as necessary to complete the picture. (From Brown (2021), used with permission.)

Findings

Analyze all Round 3 results (from your post analysis along with the input from Analyst A and Analyst B) to determine your findings. Answering the following questions will assist with your analysis.

- How do the results from each round answer the research question?

- How do the results inform the application areas and make them actionable?

Recall that the research question for this exercise is, "What is the future of the next public health crisis?" and the application area is, "What can the government (specifically research labs) do to disrupt, mitigate, or recover from the effects?" If your novel findings cannot address something that research labs should do some time over the next decade, then it is less useful as a suggestion.

- What steps in the aggregate would the Application Areas need to take?

- What are the four pivotal flags that will show the future threats are beginning to happen?

- What are the gaps in the data set? Namely:

  ◦ Who is being left out of the conversation?

  ◦ What are we not talking about?

Note that answers to the gap question, more than likely, are not found within the EBMs themselves and may need to come from additional research and other planning sessions. Notice in Figure P.7 that Analyst B chooses to pose a few questions that are left unanswered and may provide stimulus for future discussion.

CHAPTER 8

Conclusion

"With great power comes great responsibility." [13]
—Stan Lee

A Return to the Ontological Discussion

We started the first part of this book with an ontological discussion, exploring how Threatcasting was a method, a framework, and a process. At the end of this section, we return to this discussion.

Upon first reading this book readers are likely to focus on the Threatcasting Method. Being new to the method there are several phases with new activities and ways of thinking about the future. The post analysis of the raw data sets generated in the workshop is a "tough hill to climb" as a student at the lab described it.

But once the reader has climbed that hill, they can now see Threatcasting for its other benefits and attributes as a framework and process.

Framework

Now that you have climbed the hill at least once, consider how Threatcasting as a framework could be beneficial to the analyst. Moving from the rigidity of the method, how could you operate inside the phases and tools to adapt and modify?

The Threatcasting Foundation, and more specifically the Application Areas, are typically the starting point for using Threatcasting as a framework. Many analysts find they need to adjust and modify to meet the needs of the application areas. Often this happens during the workshops and "on the fly." As the analyst gets more comfortable with the Method they will feel empowered to make this shift.

Beware! It's been interesting over the years to observe when an analyst makes this shift that it doesn't really bother them, but it can unnerve the participant(s). Once the participant(s) have gotten comfortable with the boundaries and requirements of the method, they may feel like they are "breaking the rules" or "doing it wrong" when an analyst urges them to take advantage of the "wiggle room." Here the analyst will need to rely on the facilitator role to coax and encourage the participant(s).

This shift from method to framework pays off in the generation of the raw data sets. These data sets will be used in the post analysis to answer the research question and application areas.

[13] Spider-man Comic, Amazing Fantasy #15.

Generating this rich data set is so important because it will give the analyst(s) a stronger starting point in the post analysis. Ultimately producing better findings and outputs.

Process

The most recent developments of Threatcasting as a process have emerged in the post-analysis phase. Students and practitioners of the Threatcasting Method have used the three-step post analysis process and brought in new analytical tools from other methodologies.

"One methodology of interest is to analyze the data through a previously determined hypothesis. This 'hypothesis-driven' approach can compare data against an external yardstick that already exists either in the academic literature or as a practical application. This approach might be useful to control for differences between analysts that produce wildly different findings even given the same data. Because Threatcasting is a human-centric process it is susceptible to human thinking and even human error.

While thinking about Threatcasting as a process, consider that data aren't the endstate. The models and scenarios that imagine future threats help us realize our blind spots and more importantly help us discover new ways of thinking about the future. Threatcasting uses principles of analysis common in the social and natural sciences to systematically step us through a series of thinking exercises to get us beyond how we thought in the past." (Brown, 2021)

With Great Power...

Threatcasting is relatively new as a long-term strategic foresight method. As an applied method it was created and developed to be adapted. As a practitioner of the Method you now have the knowledge and the tools to envision a range of possible futures and actively work to make those futures better.

Poem

"There Will Come Soft Rains"
(War Time)

There will come soft rains and the smell of the ground,
And swallows circling with their shimmering sound;

And frogs in the pools singing at night,
And wild plum trees in tremulous white,

Robins will wear their feathery fire
Whistling their whims on a low fence-wire;

And not one will know of the war, not one
Will care at last when it is done.

Not one would mind, neither bird nor tree
If mankind perished utterly;

And Spring herself, when she woke at dawn,
Would scarcely know that we were gone.[14]

Sara Teasdale

Why This Poem Matters

Teasdale's poem has multiple meanings and significance. First, when Teasdale wrote the fiercely anti-war poem in 1918 she shows us the impermanence of humanity and how if we are not careful the world will "scarcely know that we were gone."

Second, the poem has a broader significance as it inspired a short story by Ray Bradberry (1950) by the same name. In his story he portrays a technologically advanced and automated home whose owners have been obliterated in a nuclear war. The house carries on its daily activities but with no humans involved.

Both Teasdale's poem and Bradberry's story serve as more than cautionary tales. They show us that we are not central actors in the drama of the universe. We need to be good stewards of our future or we will find ourselves without that future, overlooked, forgotten, and never known.

[14] From Teasdale, S. (1920). *Flame and Shadow.*

Part 2

The Threatcasting Method Applied

Introduction to Applied Threatcasting

With a better understanding of the fundamentals of the Threatcasting Method, Part 2 will detail the Threatcasting Method as an applied process. Threatcasting can be applied and used in multiple ways.

Just as discussed in Part 1, it starts with the Threatcasting Foundation; it is important for the analyst(s) to have completed the Threatcasting Foundation before moving further. Specifying the topic and research question will give more detail to the Application Areas. For this section of the book, specifying Who will use and How the output of the Threatcasting will be used will determine what type of Threatcasting Workshop will be required:

- Who will use the output?

- Is this for an organization, corporation, or academic institution?

- How many participant(s) will be involved?

- Will it be a large group or a smaller group of individuals?

Knowing the Application Areas will help the analyst(s) through the phases, interactions, and output of the Threatcasting performed. It will also help determine who the participant(s) will need to be.

There are three primary ways the Threatcasting Method can be applied. The Threatcasting can be a small or large group workshop or an individual activity. Each of the applications has different advantages and disadvantages. Each of the options will also give the analyst(s) varying degrees of flexibility, volume of raw data, and final output.

What will you get from the applied section of this book?

As the title implies, the chapters in the second part of this book explore the applied applications of the Threatcasting Method. This book's applied section is to be used as a "how-to" reference guide. Please note that the chapters that follow can be read traditionally, one after the other (in order to better understand application similarities and differences), or used as a reference guide for specific needs while pulling together a Threatcasting event. Feel free to move around and find the sections needed. The assumption is that you have already read Part 1; however, be prepared to reference back to Part 1 if needed.

Large, Small, or Individual?

The logistics and coordination for the large group workshop is the most complicated because of the number of participants. The small group follows a similar process but the lower number of participant(s) allows the analyst(s) more flexibility. The individual workshop, logistically, is the simplest. However, because the analyst is also the sole participant, there can be more preparation, reflection, and outside communication activities required.

For purposes of this textbook, large groups are defined as 30–60 participants and small groups are defined as 5–12 participants.

People First and Always

At the Threatcasting Lab, we take the "Care and Keeping" of the humans (i.e., core team, SMEs, participant(s), analyst(s)) very seriously. The Threatcasting process is human-centric. All humans must have their needs met to perform at their optimum along the journey.

At the lab, practicing and modeling kindness and respectfulness puts "people first." Some of the other things we do at a workshop are:

- listen and let others share;

- build in breaks and create spaces for private calls during workshops;

- whenever clarity is needed, we make our best effort to support the request immediately;

- leave space for disagreements because these are a valuable part of the process and deepen the data collected. It also builds trust and camaraderie on a team to "work through it;"

- encourage everyone involved at every step to stay open to wild possibilities and never shut things down too early on;

- encourage all voices to speak up and never hold back; and

- celebrate the ideas, laughter, and passionate sharing together as a team, and encourage all who work with us to do the same.

While Threatcasting produces serious results, humor and silliness should be encouraged during the workshops to loosen the imagination's cobwebs so ideas can flow freely.

A Threatcasting Workshop should feel relaxed and comfortable—like visiting your best friend's house—which is why we keep the coffee and water flowing freely and put candy out at every table. We also ask everyone to dress casually with our signature Lab phrase "Flip Flops Encouraged." And we build in nights out to share company with newfound friends and colleagues.

We laugh a lot, and we offer a permission space for imaginations to take flight. We are humans first and treat everyone who steps into a lab project with the same care and consideration.

You will find "stories from the Lab" sprinkled through Part 2, where we share stories of lessons learned when executing the Threatcasting Method on specific problems.

CHAPTER 9

Large Group Threatcasting Workshop

"We need radical thinking, creative ideas, and imagination."
—Mairead Corrigan, Peace Activist

9.1   INTRODUCTION

"Welcome to the Big Tent"

At the Threatcasting Lab, we refer to our large group workshops as "Big Tent" events. The name comes from a political reference. Often political platforms can be called "Big Tent" platforms meaning there is room for a wide range of people, issues, opinions, and viewpoints. We thought this perfectly describes the spirit and function of a large group Threatcasting Workshop.

A large group Threatcasting is more than just bringing people together to look at threats. The act of bringing diverse multi-discipline participants into a room feels like we'd invited the whole town—from the mayor to the teacher, from the manufacturer to the local artist. While all under one tent, we ask participants to walk through an exercise for the common good. All participants are volunteering their time and are not paid to participate in Threatcasting research. The participants' commitment to come on their own time and their own dime yields incredible results. Together, these participants develop a collection of future thinking ideas that are as expansive as possible by looking at a broad intersection of potential threats. The coordination and curation of the participants in the room takes time and thoughtful planning, but it produces a rich treasure trove of raw data for the post-analysis process.

The primary advantage of the large group is it generates quite a few Threatcasting findings. An organization or corporation uses a large group to look at possible and potential threats and align members of a group with a single vision or a single approach to the future. The output is not only the Threatcasting findings but also a team-building and team prioritization exercise.

A large group's disadvantage is more planning time is needed due to scheduling, management of space, and setting up the workshop. Additionally, there are more personalities in a larger group, and because you have so many people, everything takes more time.

Large groups participating in Threatcasting exercises are typically used for big wicked problems that require a large raw data set and many possible futures. When constructing large group Threatcasting Workshop exercises, the curation of the people in the room becomes very important as participants from diverse backgrounds, age ranges, gender, and domain expertise are vital to capturing the widest variety possible in potential futures and data generation.

Large groups are typically thirty (30) to sixty (60) participants, although they can be curated to include up to 100 participants. With a larger group of up to 100 participants, several experienced Threatcasting analyst(s) and room facilitators are necessary.

Before You Start

Before you begin this process, start with these questions.

- What problem is the organization trying to solve?

- What will be done with the results?

- What is the timeframe?

The problem or threat could modify the Threatcasting foundation or the details for how the analyst(s) conducts the workshops, so start with those questions first. If you can answer those three questions, you are ready to begin.

This chapter is designed as a "how to guide" for designing, planning, curating, and conducting a large group Threatcasting Workshop. You'll find checklists, actions, and "Stories from the Lab" to help explain the details of the process.

9.2   PHASE 0

Preparation and Curation

The Threatcasting process is built by a series of actions, each building upon each other. To complete an entire Threatcasting from beginning to end requires a plan. Here is how to build it.

9.2.1   DEVELOP THE THREATCASTING FOUNDATION

For a large group Threatcasting Workshop, the specific definition of the Threatcasting Foundation will guide the analyst(s) through each step and decision.

Pick a Topic

This problem or threat becomes the topic of the Threatcasting Workshop. Having your topic in place helps to select the language used for the invitation, workbooks, and images to be used. The topic determines the application of the methodology and the desired outcome.

Action

Refresh yourself on selecting a topic by visiting Section 2.1.1.

Research Question

Once the topic area is selected, the research question aims to narrow the focus. If the research question is too broad, unclear, or not specific enough, the Threatcasting Method will not yield the optimal results. There is a simple set of questions that can be used to help ground the research question.

Start by asking "Exploration" questions and open-ended "how" and "why" questions about the general topic. Consider the "so what" of the topic. Why does this topic matter? Why should it matter to others? Identify one or two intersecting questions that are engaging and could be explored further through research.

Next, determine and evaluate the research question. What aspect of the more general topic will analyst(s) and participant(s) explore? Is the research question clear? Is the research question focused? A research question must be specific but also leave room for complexity. Questions should never have simple yes/no answers and should require research and analysis.

Action

For examples of good and bad questions, revisit Section 2.1.1 and Exercise 2.1.

Application Areas

The output of Threatcasting identifies not only actions that can be taken by the sponsor and other parties but also events, technologies, and changes that could happen over the next decade that will indicate whether we are moving toward or away from the potential threats occurring. Now is the time to determine what the data collected during the workshop will be used for. Will the raw data be turned over, will there be a technical report, will there be articles, podcasts, or science fiction prototypes? Knowing this now will help build the foundation.

Stop Here

Don't move forward until you can clearly write your Threatcasting Foundation and articulate what the data will be used to create.

Action

Write them down to start your project documentation.

Topic

Research question:
Application Areas:

Review the topic and research question to see if they will inform the application areas. If needed, refine each of the three components to be as concise as possible.

Stories from the Lab

The important thing is to find your inspiration for the research question. Yes, there is a science to it but there is also an art to it. Some of the folks in the lab like to write a bunch of words on a white board to see how the words might connect. Others like to start with how the visualization will look on the "Save the Date" invites. It is all about how you convey to others—people you want to participate on their own dime and time—that this is an important topic and research question to explore. But, understand, inspiration also comes in the most unlikely of places and times—and you must be ready for it.

I recall in early 2017 we were trying to figure out a Threatcasting that was going to look at the intersection of Artificial Intelligence, complex automated systems, untrusted data, and a rise of nation-state power based on data/information. We had spent the day in Washington D.C. talking to various nonprofits and government agencies and had just stopped to grab dinner at a local Italian restaurant. We were sitting at a table by the window, watching the people hurry down the streets, and "it" came to us even before the wine arrived at the table. "It" was the research question for the Threatcasting Foundation and the prompt categories that we would want to explore. We were frantically grabbing for pen and paper to write it all down before the inspiration left us. I now always travel with easily accessible writing instruments.

9.2.2 ASSEMBLE THE TEAM

The Core Team

The core team includes the Threatcasting analyst(s) as well as any support or coordination staff. For a large group Threatcasting, the core team can also include the audience or users in the Threatcasting Foundation's application areas. If the Threatcasting Method is being used for a corporate, industrial, military, or governmental use, the core team could include key team members who will use and/or champion the findings. Core teams are generally between four to ten people to support a large Threatcasting.

Each large group Threatcasting Workshop has a core team that makes all the decisions (including logistical decisions) and also signs off on the final output together. The duration of the entire project can take 6–12 months.

Typically, in the lab the project length is a minimum of six months of construction time (prior to Day 1 of the event). When constructing partnership agreements between the Threatcasting team and the sponsor, one year from kick-off to deliverable works best if the time can be afforded. It is specifically true for the large group Threatcasting Workshop because with up to 60 participants (and more when you include the team and support staff), it becomes similar to coordinating a major event.

Given the potential long time-frame (from beginning to end), the core team needs to remain cognizant of any potential sources for mission or scope creep in Phase 0. Once the Threatcasting foundation is set, it can be used to keep the analyst(s), core team, steering committee, and the project in general focused. With a strong Foundation it becomes easier to hold off any potential "hijackings" or alterations to the project and workshop to refocus on short-term needs or subjects that fall outside the Foundation.

The core team will work through all the workshop details, including the topic selection, save the date invitations, subject matter experts, participants, logistics, and deliverables. This team is also essential for serving as stewards and stopgaps for problems and early reports.

Finally, it is important to acknowledge and investigate the organization's culture to determine the best communication modes, history, and politics when building a core team. Knowing these details will ensure smooth-running meetings. What does this mean exactly? It means to find out key things about the organization. For example, there are many online tools that some organizations cannot use. Knowing what is usable and what is off-limits helps with confusion and extra work.

Action

Build your core team:

Create a list of who should be included on a core team. Do this by using titles, not names. (e.g., Sponsor decision-maker, Threatcasting Coordinator, Special Advisors, Sponsor coordinator.) The Core team will be the people who will work closely with you throughout the project. Their inputs can help fill in the analyst(s)' experience or social network gaps.

1. Once you have the preferred structure of the core team, determine the individuals that you want to invite to join it. Create a list of specific names to ask.

2. Create a formal invitation to send, including frequently asked questions (FAQs) about what you are asking them to do on your core team. Sometimes a follow up call will answer any questions they might have.

3. Determine the core team meeting frequency, such as bi-weekly, and platform such as conference call, Zoom, Skype, Teams, etc.

4. Set a standardized time for each meeting. This will assist the core team members in protecting that time on their schedules. Depending on the amount of time before the start of the workshop, meetings can be bi-weekly but then will need to change to weekly as the event nears.

Steering Committee

A steering committee for a large group Threatcasting Workshop can be helpful. This is a group in addition to the core team. The steering committee does not work on logistics but instead supports the Threatcasting team by reaching out to their personal and professional networks. The steering committee brainstorms with the core team regarding the following questions.

- Is the Threatcasting Foundation sufficient?

- Who else should be on the core team and steering committee?

- What would be good prompts?

- Who would make good candidates for subject matter experts (SMEs)?

- Who would be good participants for the workshop?

- Where should the results be socialized?

The steering committee then reaches out to their vast networks to invite SMEs and participant(s). Steering committee members are chosen for each workshop uniquely. They work in a variety of domains and have a willingness to share their network with the Threatcasting team. The size of the steering committee can range from five to seven. It is important for the lead analyst to be able to manage the number of people on the steering committee. If steering committees get too large often their members' input isn't gathered, and the members can feel as if they are not being heard or informed of the project's progress. Also, an odd number helps with voting on decisions and keeping the process moving.

Action

Build your steering committee by doing the following.

1. Creating a list of who should be included on a steering committee. Note: one way to start your list is to ask the core team to open their contact lists and offer to make introductions to the individuals most connected to the research topic.

2. Writing the email you will use to the connected individuals. The focus of the email is to ask if they would be willing to meet on a group call regularly to discuss the event.

3. Writing your first agenda for the first meeting of the first meeting of the monthly steering committee. Include things like discussing dates of the event, considerations for potential SMEs, and participants' names.

Additional Analyst(s)

The lead analyst is the person who will be in charge of the large group Threatcasting Workshop. They will work with the core team to set the Threatcasting Foundation and select the prompts for the workshop. The lead analyst doesn't have to be the facilitator but often they fill that role as well. Most significantly, the lead analyst will run the post-analysis phase and be the lead author on the output.

Early in the project planning, the lead analyst and core team should consider other additional analysts. These additional analysts can support the lead in all preparation tasks. However, the main role of the additional analyst(s) is to provide a different perspective during the post-analysis and output phases. Considering the bias and expertise of the additional analyst(s) is also a form of curation, providing points of collaboration, and varying experiences. For a large group Threatcasting Workshop, you might want one to three additional analysts to help with the post-analysis and output phases.

Facilitator

Early on in the process, you should select your workshop facilitator. This could be your Lead analyst or someone that you specifically bring in to perform this function. By bringing in the individual early into the Threatcasting planning, they will have a better understanding of the Threatcasting foundation and the raw data that the analyst(s) are hoping to derive from the participants. This will help the facilitator maneuver the participants during the workshop in order to get the best results as possible.

A good facilitator can:

- feel comfortable speaking in public;

- lead a large group of people through intensive time-based exercises;

- deal with questions, conflict, and the unexpected;

- foster collaboration, discussion, debate, and even the occasional argument;

- understand the roles and policies of the organization that is hosting and sponsoring the workshop (e.g., codes of conduct and speech, safety protocols); and

- bonus: tell jokes and read poetry.

Project Documentation

As you assemble the team for the large group Threatcasting, you should also make the determination on how you plan to capture all the project documentation. Potentially either the sponsor or members of the core team might have restrictive IT processes that limit the available collaboration platforms or sharing services that can be used. Early in the project, create the data protection plan you intend to follow (i.e., backups, versioning, etc.). Further in this chapter we will discuss creating the various elements that will exist in this documentation.

Communication Plan

Building a communications plan is essential to managing the planning of a large group Threatcasting. This plan should outline how and how often information will be disseminated to the group (both with respect to the core team, the steering committee and the participants) and who will be responsible for these communications. It helps set the general expectations.

There are many communications needed to complete a Threatcasting Workshop successfully. When building a communication plan (Comms Plan), the first question to ask the core team, the steering committee, and participants is what their preferred communication mode is.

9.2.3 SELECT AND GATHER RESEARCH PROMPTS

After creating the core team and steering committee, the analyst(s) begin to gather the appropriate research prompts to address the topic and answer the research question.

Prompts

Next, comes the art of selecting prompts for the large group Threatcasting. As discussed in Chapter 2, there are many categories of prompts such as:

- social science research;

- technical research;

- cultural history;

- economic projections;

- trends; and

- data with an opinion.

How many prompts are the ideal number? The answer: enough to cover the problem space you will be exploring. Don't artificially constrain yourself at the beginning of the planning effort. Instead, remain flexible—perhaps start with four and see how that evolves.

Action

- Visit Section 2.1.3 of the methodology section for a refresher on prompts.

- Determine one core team member who is in charge of research prompt collection; allow this team member to report in at bi-weekly core team calls.

- Conduct a brainstorming session with the core team to determine the buckets of prompts that could be useful to the Threatcasting.

Subject Matter Expert (SME) Interviews

Often for a large group Threatcasting Workshop, an ideal prompt takes the form of an SME interview. An SME is a person with particular expertise, perspective, or opinion selected by the core team, steering committee, and analyst(s) to serve as a prompt for the Threatcasting Workshop. By interviewing an SME directly about the topic and research question, the analyst(s) get up-to-date, tailored prompts for the Threatcasting Workshop. The content of the SME interview is not only used as a prompt but is also used in the post-analysis phase of the Threatcasting Method.

Threatcasting Workshops work best with diverse SMEs in order to ensure the large group receives enough information to generate multiple threats. The SMEs are asked to look at specific intersections related to the sponsor's research questions and provide their thoughts as input points. They are not asked to defend their positions, just share their thoughts.

It is also ideal to have prompts that challenge and disagree with one another as this creates a richer dataset. Analyst(s) can use the SME selection as a way to bring in conflict intentionally; to do this, select SMEs that disagree with one another.

For example, if it is important for the analyst(s) to use economic prompts about the future state of a country or market, there is probably a high degree of uncertainty for what the future economic state might be. Economists are notoriously vague about what could happen. To counter this, the analyst(s) can use prompts from two different economists that completely disagree with each other. This conflict and the difference of opinion will provide the participants with a range of possible economic futures to influence their possible EBMs.

Stories from the Lab

I remember one large group Threatcasting where we started with asking the core team and steering committee for their opinions on who the interesting SMEs within the five general prompt categories that we selected to start with were. We ended up with a list of about 20 people that included Elon Musk. True—we say that the sky's the limit in our initial brainstormings. However, even with "six degrees of Kevin Bacon",[15] Elon was going to be a bit of a stretch (and honestly, would not have been helpful for the research question that we were focused on).

---

[15]   Six Degrees of Kevin Bacon is a game where players arbitrarily choose an actor and then connect them to another actor via a film that both actors have appeared in together, repeating this process to find the shortest path that leads to American actor Kevin Bacon. The game's name is a reference to "six degrees of separation," a concept which posits that any two people on Earth are six or fewer acquaintance links apart.

Ultimately, we worked with their list and added our own significantly longer list. In the end, we video-interviewed 12 individuals (8 of which we used in the Research Synthesis as prompts and 4 of which informed our post analysis). We also shifted our prompt categories during the early planning stages of the Threatcasting. So, Semper Gumby[16] should be your motto during Phase 0.

Typically, SMEs are first contacted by email to determine their interest in contributing to the Threatcasting effort. They are advised that a Threatcasting Workshop is a room of practitioners who will look at threats ten years in the future. Additional information that we include in that first email includes the following.

- As an SME, they are asked to offer 5–10 things the workshop participants should be thinking about as they model future threats within the domain.

- The SME is only given general guidance on what to talk about. For instance, they are asked what keeps them up at night regarding the topic.

- The SME is also told that the workshop participants will watch the contribution, which will be posted to a non-public viewing location during the workshop.

- The Threatcasting Lab will issue a final report with their contribution, including a transcript of their video as part of the academic documentation. However, if the SME does not want the transcript of their contribution included in the final report, then it will be omitted.

- The SME will be attributed as they designate.

If they indicate interest, the next step is to send a list of requirements. Here are some that we generally include.

- Provide the team with a 5- to 10-minute video recording focused on the research question, looking at the workshop's designated intersections.

- For the recording, excellent audio quality is vital. Therefore, using a separate microphone (if available) is preferred instead of the built-in laptop microphone.

---

[16]   Semper Gumby, or *Always Flexible*, is a take on the U.S. Marine Corps motto, Semper Fidelis (*Always Faithful*). Intended as a joke about the constantly changing dynamics of life in the military, Gumby refers to the American clay animation figure that goes on many adventures in different environments and times in history.

- No high production value is needed for the recordings. In fact, the videos are purposely "not fancy" so that participants feel more connected to the SME as a human.

- The video should be more akin to a work product than an interview.

- Many SMEs record on their laptops using the software tools they are most comfortable with for the recording.

- The SME can either record themselves or a core team member can do a Skype/Zoom call with them and record the conversation that way.

- The SME is always asked to begin the recording with their name and affiliation for the report.

Once the SME provides their work product, create a schedule to keep them in the loop. Let the SMEs know when the workshop is. Even though they will not attend, they will appreciate knowing when their materials are being used. Also, you can bring the SMEs back into the process to review the final report and contribute any additional research materials they find relevant for the report.

Action

Based on your Threatcasting Foundation:

1. Write down a list of attributes an ideal SME might have. Examples might include: Has expert knowledge of the topic, worked in the field of the topic for many years, wrote a book about the topic, is a lecturer on the topic, disagrees with another SME already identified, and so on.

2. With the help of the core team, generate a list of possible SMEs with their contact information along with a brief description for how they might work as a prompt for the Threatcasting Foundation.

3. Draft an email that explains your Threatcasting Foundation, what you hope to achieve, and ask the SME if they would be interested in adding to the Threatcasting Workshop.

4. Use your Project Documentation to track the responses.

5. Generally, SMEs prefer a prep call to explain in more detail what is expected of them. Following this call, work with the SME to record the video. Make sure to get their permission to transcribe and include it in the Final Output.

9.2.4   SELECT THE PARTICIPANTS

The core team and steering committee select the participants to take part in the Threatcasting Workshop. Some of the participants will be new to the Method and some will be previous participants in other Threatcasting events. The participants are involved in the Threatcasting process during the workshop and then during the peer review of the final findings after the analyst(s) conduct the post analysis. The participants should be sent the final report to share and have a copy of something they co-created with the team.

The curation of the room participants during a large group Threatcasting Workshop is a task that takes many months. It starts at the beginning of the conversation with the sponsor and the core team. When building the participant roster, the first question to answer is how many attendees will be in the room. This determines the capacity of the workshop venue needed. Once that number is established, then double it to create a list of invitees because aligning schedules is always difficult. If the final count for a workshop is 50 people in the room, plan on inviting 100 or more to ensure that number. Allow more than your target number to register because these practitioners have real lives, and some of them will have to back out even on the day of the workshop due to job or family obligations. If you account for that upfront, you will still hit your target number.

Project Documentation

After the number of attendees is finalized, the next step is to place names on a tracking spreadsheet as a part of your project documentation. The names are organized by domain, expertise, gender, generation, and background into a spreadsheet. Columns can also be included for links to online bios and other media. These links are an important tool when building small groups too. The spreadsheet should be reviewed on the bi-weekly core team calls to determine where gaps appear, allowing the core team and steering committee to work together to fill them. Examples of gaps could be a shortage in a domain, gender, or diversity. You are trying to model futures and threats so you can figure out how to empower people to make themselves and their surroundings more secure. To build these futures, it is ideal to invite as many diverse minds as possible to envision future scenarios.

How do you find participants?

The core team and steering committees provide names of potential participants. The team then searches each name and gathers unique data to determine their expertise and hobbies, creative practices, and other interests (these are generally from the bio links). This is important to ensure there is a diversity of imagination in the room. Imagination is an essential tool, as is creativity, when building scenarios ten years into the future.

Creating threat futures in a room with strangers demands a unique intersection of people coming together, requiring more than just different jobs. The bio links allow the room's curator to find unique humans who are as different as possible. An example of this would be learning via social media that a participant also rock climbs with a club on weekends. This rock climber then gets paired with a participant who is in a weekend rock band. They work in different fields and come from different backgrounds, creating a space for them to explore all sides of themselves and use all facets of their thinking. Hobbies and whole human activities offer glimpses of creative endeavors, and that leads to the use of imagination; these people are mandatory in the workshop. If someone is a poetry fan, they are always invited.

Outliers and Outsiders

When building the participants list, it is also important to include those that are not inside the organization, system, or industry that is being explored. Look for introductions to ethical hackers, science fiction writers, filmmakers, and poets as examples. Outsiders bring a fresh perspective to the topic and provide a unique lens into humanity.

Repeat "Threatcasters"

For a large group Threatcasting Workshop to run smoothly and efficiently, it is helpful to invite participants who have experienced a workshop before. Inviting repeat "threatcasters" can enhance the clarity of the EBM and general raw data collection because they already understand the Method and process. Ideally there would be one repeat person at each table (i.e., small working group) who will "coach" and encourage the other participants to move faster and go deeper.

Communications Plan

Generally, the communications that will need to occur with the participants are the simplest. At a minimum, they would include the following:

- a "save the date" invitation (once the date and venue are locked in);

- registration email (as soon as practical);

- logistics information (at least 2 months before the event);

- FAQ sheet (at least 1 month before the event); and

- links to the workbooks (the night before the event so that they can pre-load them on their laptop/tablet).

These elements are discussed in Section 9.2.6.

Small Working Groups

The large group Threatcasting Workshop typically takes place over two days. This time is mostly spent in small working groups of three to four participants. The analyst(s) and core team build the small groups prior to the event. These small groups are where the majority of data generation happens during a workshop.

It is vital to have dynamic small groups to create the most robust raw data and narrative outcomes. Domain and background identifiers are not enough to plan for the whole human experience. Threatcasting works because people are participating, and it is all about people. Each person is more than their job or background; they are whole humans with unique backgrounds, experiences, and interests. These individual elements are just as crucial to the threatcasting process as their job or education or what's listed on their resume. It is essential to have several bio links to review for each attendee so the team can look at each of the participant(s) through a human lens. These links might be social media posts, LinkedIn profiles, or personal and workplace websites. Many things are looked for when browsing online sites; these things may include examples such as:

- humor (shows up as the "class clown" in all photos found);

- curiosity (take a lot of online classes, shares articles about new ideas);

- creativity (is in a band on weekends);

- imagination (loves sci-fi);

- hobbies (likes to rock climb or SCUBA dive); and

- volunteer work (cares about people).

Each small group has to have these elements in addition to their domain expertise to think ten years in the future and create plausible narratives for human beings. The small group dynamic encourages members to bond, disagree, align, fall apart, come back together, get frustrated, scream, and laugh together.

The small groups offer another outcome of threatcasting—connections between people who would never have the opportunity to meet under any other circumstances. The opportunity to brainstorm with people from very different places in their professions and lives is eye-opening as a problem-solving tool. Many participants stay in touch long after the workshop. Offering a collaborative "summer camp" feeling creates the richest data because when small groups get vulnerable with each other, they are more likely to share all the thoughts they are thinking. To build a lot of raw data, you want your participants to overshare and write it all down.

Once the small groups are finalized with three or four participants, they are assigned a group name and will forevermore be identified by that name during the workshop and in the report. In the past the Lab has used colors and objects for group names: Red Pawn, Blue Pawn, Black Pawn. We have also used items such as Team Boat, Team Bulldozer, and Team Plane.

Action

Prepare your supplies for the exercise of building small groups. You can accomplish this by using the tracking spreadsheet you built, a whiteboard, or a large paper sheet. If needed, gather dry erase markers, sharpies, and phone and computer charging cables.

Just a few days before the event, build your small groups. Don't do it any sooner as some people will drop out of the workshop, and others could be added late. This activity is usually finalized the night before, or even the morning o,f the workshop. Build out your groups using your spreadsheet, whiteboard, or large paper sheet using these steps.

1. Create a grid of boxes equal in number to the numbers of small groups you will have at the workshop. As an example, 46 people would be 12 small groups of 3–4 participants each. In this example create, 12 boxes with enough space to write in the names (see Figure 9.1 as a visual example).

2. In these boxes, begin writing in participants' names by their identifiers of a domain, expertise, gender, generation, and background to ensure every team has a unique and broad makeup.

3. Assign a color to each identifier such as military/government is blue, industry is green, academia (faculty and students) is red, and so on. Create a key to see the names appear as their unique identifier visually.

4. Prepare to move names around for hours, if not days. It is so important to create dynamic small groups, so give this action a great deal of attention.

5. Once small groups are finalized, create a slide with all of the boxes, each including the small group participants' names and an assigned team name such as Team Red Pawn. This slide is used during the first-morning presentation.

Yellow Pawn	Red Pawn	Grey Pawn	Green Pawn
Sarah Valence	Leslie Grant	Karen Apple	Rachel Cash
Tim Herbert	John Cervantes	Nyle Cohen	Steve Marriott
David Dean	Ronald Currie	Pat Denton	Gary Harris
Chris Rosales	Markus McCormack	Michael Davidson	Rhyner Williams
Neon Yellow Pawn	**Bubble Gum Pink Pawn**	**Blue Pawn**	**Mint Green Pawn**
Robin Castro	Jessica Davids	Sheri Gordon	Tamara Black
Nathan Armstrong	James Underwood	Robin Coulson	Martin McCaffery
Sam McNeil	Jon-Paul Franco	Michael Tillman	John Pitt
Cary Dyer		John Spencer	Victor Goldsmith
Orange Pawn	**Pale Pink Pawn**	**Turquoise Blue Pawn**	**Black Pawn**
Elizabeth Kim	Kasandra Travis	Ashley Fuller	Whitney Lox
Wilson Lewis	Chris Hess	Damon Sullivan	Marvin Ramos
Scott Brook	Eli Nichols	Peter Marsden	Alex Buchanan
Jason Chen	Benjamin Rigby	Steven Seymour	Eric Guy
Rust Orange Pawn	**Hot Pink Pawn**	**Denim Blue Pawn**	**White Pawn**
Rhiannon Lawrence	Jessica Jordon	Samantha Miller	Heather Martin
James Christie	Madhav Prentice	Steve McKay	Josh Lees
Monty Quintero	Michael Jones	Eric Edmonds	Andrew Schultz
	Robert Britton	Michael Norman	Kathleen Sutton

Figure 9.1: Visual of participants' names and team names for a large group Threatcasting.

In our Lab, before any workshop, we obtain game pawns or tiny party favor size toys, in matching sets of four. The analyst(s) use these for the day of registration to hand out as attendees arrive, so they are aware of their small group assignment. It is a system that assists participants in identifying their small group quickly by holding their physical objects in the air and connecting with those who have a matching denim blue pawn or mini yellow bulldozer or a shoe, as an example.

The secondary purpose behind the mini objects is to give a further feel of gaming, play, and group comradery. This allows the participants to experience the two-day working session as fun. You can find pieces to use at party supply stores, craft stores, dollar stores, or potentially 3-D print them.

Figure 9.2: Visuals of team pieces that could be used in a Workshop.

Stories from the Lab

The Threatcasting Lab has hosted workshops larger than our usual 60 participant cap. We've had to make room adjustments to accommodate more people, and larger numbers cause us to get very creative with our game pieces. At a recent large gathering of nearly 70 participants, we had to purchase additional game pawns; think of the pawn from the Hasbro game Sorry! as also seen on the left side of Figure 9.2. We bought extra pawns and cans of spray paint because we ended up with 16 teams and needed colors beyond red, yellow, blue, green. In fact, we ended up with shades like bubble gum pink, neon green, and metallic rust orange. It was a great source of entertainment for the participants to have these zany colors to associate with.

9.2.5   DRAFT THE THREATCASTING WORKBOOKS

The Threatcasting Workbooks are the main tool used to capture the data that will be needed for the post analysis. The workbooks give the participants a place to write down their thoughts and perspectives on the prompts and are a place to capture the EBMs, which are the main output of the Threatcasting Workshop.

Analyst(s) should consider the Threatcasting Foundation as well as the makeup of the participants and small groups when drafting the Threatcasting Workbooks.

The collection tool can be a simple structured data gathering tool or a heavily designed (e.g., language, visual design) tool. Technically, the workbooks are a structured database, which is why simple spreadsheets work well for information capture. The analyst(s) use the workshop spreadsheets (workbooks) in the post analysis or as an input to other data processing platforms.

For ease of participants' access and version control, the Threatcasting Lab generally uses Google Sheets. Here the spreadsheet format is easy to structure the data collection and each small group has their own tab (which also allows each group to see what the other small groups are writing in real-time) as seen in Figure 9.3. Additionally, by hosting the workbook on Google Sheets, each participant can easily log into the document during the Threatcasting event.

Figure 9.3: Visual of small group tabs on the bottom of spreadsheets.

For a Threatcasting Workshop, there are two types of workbooks to build in advance. These workbooks are not given to the participants until the night before or the morning of the workshop. This is because we want them to come with their experiences, but without thinking about the problem set until they are in the room. Knowing in advance clogs the Idea Store[17] of the mind. The Idea Store has three departments: experience, memory, and imagination.

Research Synthesis Workbook (RSW)

The RSW is used with the prompts to capture the participants' analysis of the curated inputs. The RSW uses a series of open-ended questions to draw out the participants' perspectives and opinions.

Action

Build an RSW (visual shown in Figure 9.4) in your preferred spreadsheet platform.

- Use open-ended questions to build your RSW.

- The goal is to have the participants explore and discuss the prompts, processing the information.

- Typical RSW questions include:

  ○ What was a data point that you found important or interesting?

  ○ What are the implications of the data point on the threat futures or research questions?

  ○ Are these implications positive or negative?

  ○ What should we do about it?

- Highlight the cells in a color (we use a pale yellow) to indicate where participants are supposed to be entering data. It does not matter the color (just something that shades differently so that it is also obvious to any color blind participants), this will be a visual cue that they have work to do.

	Data Point #	Prompt A			
		Summary of the Data Point	Implication	Why is the implication Positive or Negative?	What should we do?
3	1				
4	2				
5	3				
6	4				
7	5				
8	6				
9	7				
10	8				
11	9				
12	10				

Figure 9.4: Visual of blank RSW.

---

[17] The "Idea Store" is a term coined by R. L. Stine (prolific American author of over 330 books). https://www.masterclass.com/classes/rl-stine-teaches-writing-for-young-audiences/chapters/the-idea-store.

Rarely are these questions changed as they will apply to most workshops and situations. For a refresher on the Method behind building an RSW, visit Section 3.1.2.

Threatcasting Workbook (TCW)

The second workbook to build for the workshop participants is the TCW, which contains a series of questions and tasks that lead the participants through the design and modeling process to develop their EBM. This process uses the tools of Experience Design and allows the participants to explore a person, in a place, experiencing a threat—all based upon the prompts and the RSW information.

Action

Build a TCW in your preferred spreadsheet platform. For a refresher on the Method behind building a TCW, visit Chapters 4 and 5 (which include visual representations of these concepts).

The TCW should contain the following.

1. Places for the participants to include their team name and a place to name their future.

2. Space to include the selected data points from the RSW.

3. Person Questions provide space for the participants to describe their person in a place experiencing the threat.

   a. Consider what extra questions might help the participants tell a richer and expanded story. What would fill out the vision and give the analyst(s) more information for the post analysis? Use the Threatcasting Foundation. The research question and application areas can give specific guidance and language for more explanation.

4. Add in Experience Questions using the experience design approach.

   a. Again, refer to the Threatcasting Foundation's research question and application areas as an initial prompt for where participants might give additional detail.

   b. Use The Six Dimensions of Experience Design to explore different aspects of the experience questions. (Breadth, Duration, Interaction, Intensity, Design, and Value)

5. Add in enabling questions to explore what factors helped bring the threat about using the EBO approach.

a. The Threatcasting Foundation's application areas will give analyst(s) areas to consider for more questions for the workbook. These questions should highlight various enabling areas or activities.

    i.   What economics or businesses need to be in place?

    ii. Is there new research that needs to be conducted?

    iii. Is there a specific failure of policy, law, or culture that enabled the threat?

    iv. Are there geopolitical conditions that contributed?

b. Each of these enabling areas can help the analyst(s) expand and specify areas where they might want the participants to comment.

6. TAD Backcasting: Add in the backcasting space for gates, flags, and milestones.

a. Are there specific questions you can ask that will help give the Who in your Application Areas more specific actions to take when they apply the findings?

9.2.6   LOGISTICS

The logistics for conducting a large group Threatcasting Workshop can be complex and unique because of the number of participants. The following is a set of tasks to help the analyst(s) and core team conduct a large group Threatcasting Workshop.

Plan your kick-off

The kick-off for a Threatcasting Workshop happens when the core team decides to run a workshop to gather the raw data to determine the Threatcasting Foundation. Often this is "sponsored" by the person or organization that will be utilizing the findings as defined in the Application Areas of the Threatcasting Foundation.

    Planning meetings are generally all done over conference calls and/or video-calling platforms. For each large group Threatcasting Workshop event, the analyst(s) and core committee will want to schedule bi-weekly calls for the first six to eight months, then weekly calls the month before the event.

Action

For the calls:

1. Select a team member to be the call coordinator. This person will email the link to the conference call number to bring everyone together.

2. Set the agenda each week to move the project forward and bring everyone together.

3. Send out the agenda for the call 24–48 hours prior.

4. Select a team member to be the meeting host or facilitator to keep the topics moving.

5. Select a team member to be the scribe or note taker for the call to capture the discussion as well as the due outs.[18]

6. Follow up with any due outs after the meeting within 24–48 hours.

Pick a Date

The top agenda item is to determine the date for the Threatcasting Workshop at the first kick-off meeting. Once that is established, everything else can move forward. That seems like a simple action, to pick the date, but it never turns out as planned. Typically, the date is chosen at least six to nine months out because many participants will have packed schedules and need that much lead-up time to clear two full days (plus any travel time associated with the location).

It's all about the people, so selecting the right date so the right people can be there is essential. Plan not to schedule near important industry-related events or conferences. Additionally, try to avoid scheduling around holidays, but also avoid the standard Spring Break timeline as many families might already have travel plans, additionally hotel accommodations near the Threatcasting site might be more expensive due to other travelers. In the Lab, we often schedule workshops in the spring, encouraging participants to travel from colder climates to sunny Arizona. Consider the day of the week carefully as well when determining dates. For example, consider leaving Monday as a travel day so that participants are not required to travel on the weekend and leave their families. This is because many organizations frown on approving travel on the weekends as that could result in comp time (if an organization is supporting the traveler). Having the date upfront also makes it easier to reserve the facilities and gives you enough time to plan all the actions.

Location Selection

As soon as the core team selects the date, the workshop's location should be selected. For planning purposes, the core team must also come to a quick decision on the general split of participants (and their home locations) in order to decide on the best location for the Threatcasting Workshop.

Action

Write down answers to the following questions regarding the selection of the location.

---

[18] DUE OUTS are all the things that need to get done before the next call. These tasks should always have a name attached to them so that individuals can report in on progress at the next meeting.

1. Is travel possible for all attendees?

2. What is the travel time and distance for most participants?

3. What time of year is it? This matters because those in cold climates would almost always prefer to take two to three days in a warmer climate as opposed to vice versa. The time of year can also influence hotel costs.

4. Beware of locations with a negative connotation for some of your participants. For example, government employees have a hard time justifying attendance at events in Las Vegas based on the nature of the city.

5. Pay specific attention to any location restrictions that the Threatcasting sponsor has levied on them by their organization.

Venue Selection

Once the location is selected, it is important to select the venue in that town. For planning purposes, the core team must also come to a quick decision on an approximate size of the group, as this will determine the venue required.

Consider the following questions when analyzing potential workshop venues.

1. Does the venue have enough space to host the number of participants you plan to attend the workshop? Consider that you will need adequate space so that everyone can gather together as a large group for portions of the event and have enough space to break into small groups for a portion of the event.

2. Is there room for the team and staff supplies and gear? Generally, look for a separate side room that you can store "things" that you might need during the course of the event.

3. Is there a side room participants can duck into for calls? Life never completely stops for a participant and sometimes they will have to take that work call. Therefore, having a semi-private space they can duck into is something nice to have.

4. Is there an overflow space where participants can spread out during their small group sessions if they need a change of scenery?

5. Is there space to provide catering? Generally, you will need a couple of long tables that can hold the day-long catering offerings. Preferably this should be in the main space or in a location that is generally accessible by the participants and not random people.

6. Is Wi-Fi easily accessible for all participants? Or will that be an additional charge?

7. Are there enough power outlets at each work area for the small groups? Assume that each participant will want to plug in their phone and their laptop/tablet.

8. Is there audio-visual equipment available in the main Threatcasting area? You will want the ability to project the facilitator's laptop with audio capability for the presentation of the prompts.

9. What are the costs of the facility rental and how far out can dates be locked in?

10. Does the venue allow outside catering or will you need to use one of their preferred vendors?

11. Are there any restrictions to the time the facility is available? Namely, you want to make sure that you can access the facility in the morning at least an hour before the start of the event and host the event, un-interrupted, during the day. Generally, we try to reserve facilities from 7:00 a.m. to 5:00 p.m. on the event days.

Generally, the event space needs to have the following:

- enough desks/chairs in the main room to facilitate arranging the tables in small groups;

- couches or other table and chair seating, so small groups can move outside of the main room if desired;

- screen, projector, and sound system in the main room;

- IT support available;

- a nearby small conference room in case participants need to take a private call during the workshop;

- plenty of parking or easy transportation options nearby;

- food options nearby for pick up or delivery;

- a reasonable and affordable price within the event budget; and

- free Wi-Fi and/or included in the space rental fee.

Action

1. Contact rental venues to see what kind of packages they offer to get a sense of what is available in the area where the workshop will be held.

a. Ask the core team and steering committee if they have ideas on viable event locations (and they might have the ability for a discounted rate).

2. Create a tab on your tracking spreadsheet called venue options. List the name of the venue, contact number, website, contact person, and link to the event package. This tab can assist with venue selection during core team bi-weekly meetings.

3. If possible, coordinate a time to walk through the potential venues taking pictures in order to share with core team members.

Build a Save-The-Date Announcement

Once the date for a workshop is determined, a "Save the Date" digital postcard should be created. This digital postcard is sent to possible participants three to six months in advance.

Figure 9.5: Save-the-Date Announcement (large group Threatcasting Workshop).

The graphics on the save-the-date postcard should relate to the Threatcasting Foundation for the workshop. An example might be if the workshop is focused on space, using images of satellites and stars would work well. The digital postcard should include the workshop title, location, date, and research question. An example from the Threatcasting Lab is Figure 9.5.

Action

1. Assign one core team member to be in charge of all communication outreach. This individual should send the digital save-the-date postcard once participants are identified.

2. Create the digital save-the-date postcard including these items predetermined by the core team:

    a. Workshop title.

    b. Workshop location.

    c. Workshop date(s).

    d. Workshop's research question.

    e. Imagery related to the workshop's focus.

3. Be cognizant of the file size of the final image. Many email servers will block emails with large attachments and your recipient will not get the email. It can help to convert the file and export it as a .pdf as this will automatically reduce the file size.

Stories from the Lab

"Why should I come to the Threatcasting Workshop?" an executive at a large design company asked. This happens sometimes, especially when people have never been to a Threatcasting Workshop before. It's understandable. People are busy and their time is very important.

"I think you should come to the workshop because you care about the threats we are going to explore and you also care about national and economic security," I replied with an encouraging smile.

"Yes, yes of course I do," the executive replied. They looked a little nervous. "But what do I get out of it?"

"Well aside from helping make the country and the economy more secure…" I began. "You also will spend two days with people who are your peers and who are passionate about the same subject. You will work closely with them and get to know people you would never meet before."

The executive didn't look convinced. "But what will I take away?"

I had had enough. "If national and economic security and collaborating with passionate like-minded people isn't enough, then I'm not sure we want you to come to the workshop."

I smiled. They smiled back.

"Ok, yeah you are right." they replied. "I'll come. I just needed to ask."

***

It is impossible to convince a person to do anything. This includes getting involved in a workshop. All you can do is lay out the benefits, the subject, and the excitement. It's a hard lesson to learn. But for some people it may not be the right time for them to come to your workshop. You will want people to attend because they will devote their full energy to the workshop, which is a lot of work.

If you can't convince a person to come, don't get discouraged. Likely it wasn't the right time for them and they would not have been a good participant. With a strong and engaging Threatcasting Foundation expressed simply, the right participant will engage with your workshop.

Registration

When practical after the "save the date" announcement, a registration email should be sent to the desired participants. This can either be sent from the core team member in charge of outreach or can be sent directly from the event registration site.

There are many online event registration sites to choose from and the core team should select the site early in the planning stages. Additionally, you should always check with the Threatcasting sponsor to ensure the selected registration platform is acceptable for their organization. Also, knowing if the selected registration platform routinely has their emails routed to spam within targeted organizations is important to find out.

The registration platform should, at a minimum, be able to collect the following information about the participants:

- full name;

- organization;

- preferred email to contact;

- confirm that they can attend both full days of the Threatcasting Workshop; and

- food allergies/restrictions.

Action

These are the steps to take for workshop registration.

1.   Select which event registration site to use.

2. Create an event with relevant information about the workshop. Ensure that you either create it as a closed event (so that only people you designate can register) and/or include an acceptance step for their registration to be complete.

3. Test out the site's functionality with core team members before opening it up to the desired participants.

4. As soon as the individual responds with interest to the "save the date," send another email letting them know they will receive an invitation to register from the registration site and note the site's name so they know what to search for in case their SPAM filter grabs it.

5. If one of the team or committee members states someone is attending, check the registration list often and send emails to those who are halfway confirmed by your team or committee to ask if they need assistance completing registration.

Using an online registration site makes tracking all participants easier. The site will allow reports to be run for the bi-weekly calls.

Logistics Information

General logistics information should be sent to the confirmed attendees as soon as possible. This will aid them in booking their travel plans and lodging accommodations if they are traveling from outside of the local area. More information about planning accommodations is located further in this chapter.

At a minimum, you should consider including the following information:

- address of the Threatcasting Workshop venue;

- public transportation options to arrive at the venue;

- closest parking garage options (with cost if available) to the venue;

- local lodging accommodations information (especially if a hotel block has been saved at a discounted rate for Threatcasting participants);

- closest airports to the venue for out-of-state participants; and

- rough schedule for the two days (so that participants can plan their travel accordingly).

FAQs

The registration should include an FAQs list so the participants know what is required of them. The FAQ list can be sent as the follow registration confirmation. Having these details allows the participants to show up relaxed and ready to step ten years into the future.

Action

Build an FAQ, which could include some of the following items to help address the culture and execution of the event.

1. Participants should dress comfortably and casually. The Threatcasting Lab's signature line is "flip-flops encouraged" because we want everyone comfortable for the two days.

    a. Please do not wear a suit or a uniform.

2. You will do a mixture of large group and small group work. Small teams of three to four people are pre-assigned by the Threatcasting team. The assignments are provided once all participants are in the room.

    a. No, you cannot find out who is on your team before the day of.

    b. No, you cannot request to be on a specific team.

3. Teams stay together and work together during the entire two-day event—everyone works; no one observes.

    a. No, you cannot visit and watch the process. This sausage-making takes all hands on deck.

4. There will be breaks for phone calls to do a quick check-in.

    a. Your small team needs you, so do not plan long conference calls during the workshop.

5. This is a working event; please bring a personal laptop that can access Wi-Fi and access the sharing platform that we are hosting the workbooks on.

    a. You will receive the links the day before the workshop.

Stories from the Lab

It's hard to get mad at someone when you can see their toes.

Encouraging participants to dress comfortably and casually is a key to success. All uniforms and suits are prohibited within the workshop space. This is not a small ask for participants coming from the military where uniforms are mostly worn or from corporate America where suits are the standard. On Day One, you will still have participants arrive in dress pants and dress shirts, but after seeing the permission space afforded at the Threatcasting Workshop, they normally come around for Day Two and show up in shorts and a polo shirt.

Why this dress code? We have discovered that if people are in more comfortable clothes, the working sessions are more comfortable. The other reason is that we have all ages and ranks designed into our small groups, and it is easier for a 20-year-old student to speak to a 4-star general when they are both in jeans and a casual polo. It levels the playing field.

The Threatcasting Lab's signature line is "Flip-flops encouraged." To embrace this culture, normally the facilitator wears sandals or flip-flops on the first day (and in the winter, it is socks with sandals). Additionally, it is important for all the team members involved in the execution of the event (i.e., facilitator, support staff, analyst(s), core team, etc.) to model the dress code on Day One.

Accommodations

As many participants will travel from out of town, it is important to select an event hotel near the workshop venue. The partnership with the hotel should be arranged early in the process, right after the date is agreed upon. Some hotels allow for the event group to block a number of rooms, often with a discounted rate—it's always worth the ask. This is especially true if attendees include military and government employees as they will require specific rates per their departments. Government per diem rates are located at: https://www.gsa.gov/travel/plan-book/per-diem-rates.

Having all of your attendees stay at the same hotel promotes after-hours social experiences, and this encourages conversations to continue.

Travel and Transportation

As a courtesy, it is nice to provide attendees with preferred airline information months before the workshop, if applicable. Provide the distance from the airport or train station to the workshop venue and a list of all possible ground transportation, including rideshares, cabs, light-rail trains, bike, and scooter rentals. Include this information in the FAQs and/or logistics information email as discussed previously in this chapter.

Project Documentation

The established communications plan will be useful when scheduling bi-weekly meetings, sharing notes, and creating outcomes for the core team. Many planning documents are needed for every large group Threatcasting workshop, so before building the project documentation, the team must designate which online sharing platform will be used to host the shared documents. Some organizations have restrictions on what individuals can access from their work computers.

Action

Create a spreadsheet on a sharing platform such as Excel, Teams, or Google Sheets with tabs as the following.

1. Potential SMEs Tab with columns to denote name, affiliation, gender, domain, background, bio links, and status such as accepted, recorded, transcribed.

2. Potential Attendee Tab with columns to indicate name, affiliation, gender, domain, background, bio links, and status such as invited, registered, and attending.

3. Seat Matrix Tab, which is used to create small groups and ensure the diversity of attendees. Columns denote name, affiliation, gender, domain, background, bio links, previous attendee, and recommender's name (see Figure 9.6).

	A	B	C	D GOV	E ACAD	F IND	G MIL	H Gender	I Speciality	J Email	K Link to Bio or Profile	L Recommended by
	First Name	Last Name	Organization									
	Sarah	Jones	US Army				x	F	strategic planner	xxxx@xxxx		
	Tim	Martin	MIT		x			M	PhD in AI	xxxx@xxxx		
	John	Moses	Target			x		M	CFO	xxxx@xxxx		
	Allie	Smith	Treasury Dept	x			x	M	financial crimes analyst; previous Army Ranger	xxxx@xxxx		

Figure 9.6: Threatcasting Seat Matrix.

4. Team "TO DO" Tab is a spreadsheet to track bi-weekly due-out tasks (see Figure 9.7). Many of these will be logistical tasks dealing with the venue and the "care and feeding" of the participants.

A Rank	B Item	C Status	D Next Steps	E Owner	F Links
1	Venue	Tentatively Booked	Review contract; Sign contract; Send deposit	Jane Smith	http://location_of_venue
2	Core Team Meeting	April Meeting Complete	Send calendar invite for May invite	John Rose	
3					

Figure 9.7: Threatcasting TO DO tab.

5. Venue Options Tab is a listing of all the venues and the assorted information that you have researched within the city you plan to host the Threatcasting Workshop.

Stories from the Lab

Dice Thieves

In the Lab we typically use dice as a tool and method for participants to quickly pick their data points from the RSW (see Section 9.4.1 for details and other methods). So, this is something else to collect during your preparation for the workshop.

We use dice (Figure 9.8) because using them moves things along and has a low level of randomness that is not science or math, but is FUN! That's something to remember when it comes to the group work: It can actually be fun. People like to roll dice. Most people in their daily lives don't get to roll dice all that often. It's novel for them and a great conversation starter.

But be warned!

People will try to steal your dice. It's funny how grown, law-abiding adults can quickly turn into petty thieves when they start rolling dice. It's almost as if the dice roll is a slippery slope to vice and crime.

We have made it a joke and we quickly police participants, "keeping an eye on them." This gives the team a sense of shared vice and helps with group cohesion. It also gives the facilitator another way to interact with the groups in a more relaxed way.

On more than one occasion we have received ransom notes from participants who have taken dice and want a cookie for its return (true story!).

Figure 9.8: Visual of dice.

Note: If you are running a virtual workshop via a video platform, many of the logistical items will not be needed. That said, if you are attempting a virtual workshop with more than 30 people, we'd like to hear about it because that will be hard.

Planning a No-Host After-Hours Gathering

At the end of the first day, we typically schedule a no-host after-hours gathering at a restaurant within walking distance from the Threatcasting venue. The workshop is done for the day, but the participants want to continue to talk and connect. Invite them to do exactly that by pre-arranging with a venue to host the number of individuals you will have in attendance, explain you are not covering the costs, and each person will pay for their own meal and drinks. Confirm this is okay with the venue and continue your night with great conversations and connections.

This opportunity creates more of the "summer camp" feeling you want to create, and when participants come back the next day, their trust in each other has grown, and they do a deeper data mining of themselves and each other.

Stories from the Lab

Name Tags—Threatcasting Version

Step one to name tags is to purchase or make creative name tags. This is the first in-person engagement with participants, so allow them to step into play and creativity immediately. We like to use sticker name tags that have comic book graphics around the outer border.

Next, each name tag should have only first names, and they should be handwritten; this immediately signals a human touch and fun. First names (and a lack of organization titles and affiliations) help set every participant as an equal member in the group. Recall each participant was hand-selected to participate in the Threatcasting Workshop because of what they bring to the table.

Create name tags specific to the small group teams, use markers to match colors, and select any borders on the tags to match by team. These efforts tie the small groups together, further bonding them and creating trust and vulnerability because they feel like they are a part of a team. Because they are!

Build Playlists

You will need two types of playlists for threatcasting. The first is to play in the main room when the small groups are working. Having a playlist such as jazz or symphony on low volume creates a calmer, more creative environment. The second playlist is for when the analyst(s) comb through the pages of raw data post-workshop.

Prep Equipment

The workshop room should have a drop-down screen (or adequate large television monitors) to present the workshop PowerPoint and the SME videos. The room must be Wi-Fi equipped. The core team should have an IT person on stand-by, as participants will have trouble logging in and accessing the workbooks—every single time, every single workshop. The logistics team laptop is in the room with all documents and presentations pre-loaded. Electrical outlets, extension cords, charging towers, and multi-platform chargers for phones and laptops are often requested, so it is better to plan ahead and have them already on hand. Plan to have a phone app or an actual stop-watch to keep time during the report outs.

Order Catering

Catering should be arranged months in advance of the workshop to ensure paperwork and down payments are complete allowing for no surprises the day of the workshop. Registration (as stated previously) should have a fill in the blank option to ask about food restrictions, allergies, and special requests. An email should be sent to all registered attendees one week before the caterer's finalization date. This email asks participants to confirm they are attending and if they have food restrictions. This final count is generally provided one week in advance of the workshop to the event caterer as a headcount and restrictions requirements.

Catering will require tables to set up on and possibly electricity. Each morning of the workshop, coffee, tea, and water are provided to keep participants focused and from missing important opening information because they need to stop by the coffee shop. It isn't always possible to offer lattes, but black coffee with cream and sugar keeps most participants happy. Water stations should be available daily and frequently refilled.

In order to maximize the time with the participants, the small teams work through the lunch hour, so providing boxed lunches with to-go drinks is generally the most convenient option. This saves a considerable amount of time as participants do not have to leave the venue, allowing them to stay in the zone working together. Most participants love this idea of working lunches as it demonstrates that we value their time (and sacrifice to attend) and we want to maximize their contribution to the effort.

In the afternoons, participants should be offered healthy snacks such as trail mix to keep going until the end of the day. Catering also means trash and clean up after each food service. Having a team member with some experience in this area helps keep things running smoothly and behind the scenes so participants can focus on their small group work.

Often the event space will have restrictions on the types of caterers that can be used within their facility. Therefore, make sure this is part of the decision-making process with the core team when it comes to choosing a venue.

For the benefit of happy participants, keep the coffee, tea, and water flowing all day long. Having a station set up allows participants to stand up, step away, and loosen their minds as they refill.

Bowls of candy on every table are greatly appreciated by the participants—anything from individually packaged mints to mini chocolates to lifesavers. Assign someone on the core team to obtain candy bowls and candy for both days of the workshop.

Check-In

If possible, access the event space the evening before the workshop to ensure set-up details (such as tables and chairs, projectors, screens, and space for catering) are complete. This way there won't be as large of a scramble in the morning.

This is our morning of checklist.

- Place a table for volunteers/team members to sit where they can greet participants as they arrive. This should be right outside the event space, clearly visible from the entrance to the floor.

- Place the final registration list on a clipboard (bring pens), laptop, or tablet (don't forget the charger and stylist) to check participants off the list as they arrive.

- Sort participants' name tags in alphabetical order on the table surface for quick pick up as the participants check-in.

- Hand participants their name tag and small group assignments. These should be in the form of colored game pawns or mini plastic toys. Ask them to hold on to the objects for the morning.

- Let participants know that for the opening morning session they can sit anywhere in the room they want to. They do not need to find their groups yet.

- Inside the workshop room, tables and chairs should be pre-set for teams of up to four.

- Catering should also have a table inside the workshop space or conveniently nearby. For the morning, it should be ready with coffee, tea, and water.

- The screens should be ready for viewing, all presentations loaded, and wi-fi turned on.

- Extension cords and charging stations should be placed throughout the room.

- Place bowls of candy on every table.

9.3   PHASE 1

In Part 1 of this textbook, we explained Phase 1 as the Research Synthesis portion of the Threat-casting Method. However, when discussing the application of the Threatcasting Method, you should visualize Phase 1 beginning at the start of Day 1 of the Threatcasting Workshop.

9.3.1   WORKSHOP OPENING

Because of the number of participants and the logistics of the event, the opening of the large group Threatcasting Workshop can be quite the event.

It starts with greeting the participants into the venue space to set the stage for the innovation and critical thinking that will occur during the two-day event—which, like at any large event, can turn into herding cats at the start to get everyone in their seats and ready to begin.

> **Stories from the Lab**
>
> Opening Morning
>
> *At the Threatcasting Lab, we believe words matter greatly, and this is why we open every workshop with a poem. Our team takes turns selecting a poem and reading it at the opening as our welcome to the workshop. We also cite these poems in our reports because, to us, humanity should always come first. Open your workshop with a poem to invite a more human experience.*

This is our general opening session checklist.

- Selected team member reads their chosen poem as the opener.

- Selected team member welcome everyone and reviews housekeeping items such as bathroom locations, fire exits, etc.

- Selected team member gives a brief history of the lab.

Workshop Agenda

Each day of the workshop should have an agenda built as a slide deck to provide instructions for the participants. The Threatcasting Lab's deck (for Day 1) includes the following items:

- introduction to the Threatcasting Process;

- relevant definitions within the problem space;

- reveal of the assigned Small-Group Teams;

- Phase 1: Research Synthesis; and

- Phase 2 + 3: Futurecasting + Backcasting.

Let's say that you want to start Day 1 at 9:00 a.m. Then you should "schedule" registration and coffee from 8:30 a.m. to 9:00 a.m. to attempt to get the stragglers in before the event starts. Generally, the opening session takes about 15 minutes. Then we spend about 45 minutes reviewing the agenda and discussing the Threatcasting process (recall that for many of the participants this will be new for them). Additionally, due to the diverse backgrounds of the participants it is important to review critical definitions within the problem space the Threatcasting workshop will tackle.

For example, we conducted a Threatcasting on the intersection of weapons of mass destruction (WMDs) and cyber. To meet the required application areas for the sponsor, it was important to clearly define what we meant by WMDs and cyber (giving both the international law definitions, the U.S. military definitions, and Websters' Dictionary).

Once that is complete, we reveal the small group teams so participants can see who they will be working with for the duration of the workshop. Then we dive into Phases 1, 2, and 3 of the Threatcasting Method. Viewing the prompts for Phase 1 will generally take the remainder of the morning. This allows the small groups to conduct their analysis during lunch and report outs can occur in the early afternoon. Generally, we shorten the time to execute Phase 2 and Phase 3 as we really want participants to get comfortable with the concepts on Day 1. This, generally, fills Day 1.

Facilitator

Once the opening session concludes, the floor is yielded to the workshop facilitator to start the show. The facilitator is in charge of keeping the room's time, keeping the participants on task, and making the experience feel like a fun and creative workshop. A facilitator should have a dynamic personality and a boldness that allows for maintaining control of the floor. Because of the number of people, facilitation of the workshop will help move the event along and help the quality of the raw data.

Given that this is a large group Threatcasting workshop, you should plan to have at least one co-facilitator to assist. If the group is toward the high end (i.e., close to 60 participants), then you should have two co-facilitators to assist with the myriad things that will pop up during the event (both Threatcasting related and logistics related). These facilitators could come from the core team or might be volunteer support staff.

Throughout the entire workshop, the facilitator should frequently ask participants if they need anything or have any questions; they will, and walking around, continuously connecting with them makes the process easier for all involved. The facilitator should frequently remind the participants that they are experts, and that's why they were invited to the workshop.

Workshop Presentation

Build the presentation for the workshop in advance. Often PowerPoint can be used. It's important to tell the participants what we're going to say during the facilitated opening and then say it, and then again as a way to over-communicate. And then on Day 2, tell them again what you already said. This helps set the expectations, and the repeated messages build familiarity and trust with the participants. We keep the PowerPoint presentation up on the screen during the entire day. This way we can either highlight the current directions for the task at hand or display the current timeline.

9.3.2   PRESENTATION OF PROMPTS

Back to the "art" of conducting a Threatcasting Workshop, it is critical to think about the order of presentation of the prompts to the participants. For instance, should you present the technology prompt before the cultural prompt, etc. Consider the message that each prompt is conveying and how each might add or subtract from the other prompts. Also, consider the length of the video recordings for each prompt—mixing up the long and short prompts is better for the participants' attention span. You will also need to think about building in a bathroom break in the middle of the prompts so that people can disengage from the medium for a few minutes.

Once the prompts are ordered to your satisfaction, then you will need to assign small groups to each prompt. The general concept is that everyone watches every prompt so that they have a breath of understanding of the problem space. However, each small group is assigned one prompt that they will pay specific attention to and ultimately, conduct the research synthesis exercise on.

Selecting which group will be assigned to which prompt should also be a deliberate action. Consider some of these ideas when assigning.

- Look at the composition of each group and attempt to assign the group to a SME prompt that will generally be outside of their general wheelhouse. This will force the group to think about new subjects.

- Alternatively, you may have a group that you specifically want to review a particular prompt because that group has an expertise in it. The group's expertise will give the larger group more detail and fuel once the RSWs are combined and the groups begin to use them for their TCW.

- There might be a group that is particularly interested in an SME and their subject area (e.g., a new technology, an interesting piece of social science, etc.). You can assign that prompt to them because it allows them to explore the new data with an enthusiasm from which the entire group will benefit later in the phases.

- inally, there is not a 100% right way to assign the SMEs and groups. But it is essential that the analyst(s) have a logic behind why they were assigned. This should be included in the project documentation for use during the post analysis. Reviewing why the group was assigned the SME before reviewing the results in their RSW can give an important perspective to the data in the workbook.

Given that the groups are generally crafted the day before the workshop starts, this assigning of groups to prompts tends to happen the morning of the event (especially once we see whom shows up and which groups have to be shuffled due to "no shows"). Given that this is a large group Threatcasting workshop, you could normally have two to three small groups focused on each prompt. This helps with the richness of the data set created, resulting in stronger EBMs created.

The breakdown of which small groups are assigned to which prompts are displayed on the powerpoint presentation at the front of the room. Make sure to allow individuals to identify which prompt they will need to take notes on. Then start playing the prompts for the participants. We show the assignment of prompts to groups between each prompt playing as a reminder to the groups that need to pay the closest attention to the next prompt.

Once all the prompts are presented to the participants, they are encouraged to find their small group team-mates to start the data capture in the RSW. Depending on when the day starts, we tend to incorporate the box lunch into the small groups' time to conduct research synthesis on their prompt.

9.3.3 DATA CAPTURE

At the start of each Threatcasting workshop, all participants are provided with a link to the workbooks. The small teams will use these to write their ideas, notes, and scenarios. Using workbooks allows more data to be captured than appears in the verbal report outs because no one can do justice to everything that was in their heads. The idea then for the workbooks is to get as much detail written down as possible. The workbook gives the small group space to start looking at the problem from various angles. As the workshop progresses, each small group begins to populate the workbooks' fields.

Once the small group gets together, we encourage each individual to share what they individually captured during the prompt presentation as the important points when it comes to modeling the future. Then they collaborate together to create the final list of points which can also include a sum of their experiences and expertise overlaying what the SMEs provided. They should record their work in their small group tab (labeled with their team name) within the shared RSW.

One of the goals of the RSW is to capture the "wisdom of the room." The facilitator should encourage each member of the group to give their specific take on a data point. Success during the RSW is to capture what each member of the group thinks about the prompt.

One helpful tip is to encourage people to disagree with the SME they are discussing. This technique often frees people's opinion when they understand that there is no "right" answer. The goal is to capture their thoughts.

It is important for the facilitator to keep an eye on how the workbooks are being filled out to ensure everyone is typing in their appropriate tab. Every time and in every workshop, multiple groups will start typing in the wrong tab and while funny to watch, it is important to head off as soon as noticed.

In a large group Threatcasting, you will have multiple small groups conducting research synthesis on the same SME prompt, which will result in some interesting and unique ideas. The more information the groups can provide at this stage, the smoother Phases 2 and 3 are.

A Note on Scribes

When the small groups gather to put information into the workbooks (RSW and TWC) the facilitator should encourage the small group to pick a scribe. This is the person who will capture the conversation, discussion, and disagreement that happens in the group. The scribe is essential to the Threatcasting Method. If the data is not captured fully in the workbooks, it cannot be used later in the workshop nor will it be of any use to the analyst(s) during the post analysis.

Because the scribe is so important the facilitator should check in with each group and identify the scribe. They should be thanked and told that they are an important part of the method. The facilitator should then encourage the rest of the group to also write in the workbook, to capture as much information as possible. High performing groups will often have multiple members of the small groups adding information to the workbooks.

For a workshop that might take place over multiple days, the facilitator should encourage the small group members to share the scribe task (and switch it up). A single person should not be the scribe for every step in the process if possible.

9.3.4   DISCUSSION

Throughout the course of the Threatcasting workshop, the small groups will "report out" multiple times to the other participants in the room. Small groups report out so that everyone in the room can hear what everybody else is thinking. During report outs, participants have a strict time limit to present the most important thing(s) that came from their previous activity. With respect to this first "report out," the small groups are typically asked to provide the top three important data points that came from their research synthesis activity, including the answers to the other research synthesis questions for those three points. The challenge is that they aren't allowed to repeat anything that a previous small group stated for their SME prompt. This provides more of a challenge to the second and third small group for a specific SME prompt than the first small group that gets to go. The report outs should generally occur in the prompt presentation order (all small groups on prompt #1, followed by all small groups on prompt #2, etc.). The facilitator should randomly pick the order of the small groups. The facilitator also needs a stopwatch to keep track of the time.

When the timer goes off at the end of the allocated time (perhaps 2 minutes), we let the alarm go off in the room, so the reporter knows time is really up. We have had participants run from us and try to continue to talk over the alarm. This requires a strong facilitator to cut their reporting at the two-minute mark. Yes, they might have more to say but it also allows the other groups to speak and share their insights on the prompts. This will also, generally, spark side conversations at the break.

Once all the small groups have contributed, you could open the floor to short conversation to encourage participants to share what didn't get talked about; perhaps highlight some of the side conversations they had in their small groups that were not presented. "What didn't come up" is an important element of Threatcasting because what isn't talked about often illuminates the things that should be discussed. You should have an analyst(s) or a co-facilitator take notes on this conversation as the comments might be helpful in the post-analysis phase.

9.3.5   PREPARE FOR NEXT THE PHASE

When creating the agenda for the day, time needs to be allocated for the magic transition of the data participants provided in the RSW to the set of final data points that will be used during the rest of the Method. For example, if three small groups conducted research synthesis on prompt 1, that information needs to be consolidated.

While the facilitator has the attention of the participants in explaining the next activity on the agenda, a co-facilitator can be conducting data management on the RSW. Generally, we create new tabs within the RSW that have the consolidated listing of data points for each prompt. Using the same workbook (which we know participants can already access) makes it easier for access. It also highlights the transparency so that participants can trace back, if desired, which small group made the original points.

For purposes of illustration in Figure 7.1, Slot A will hold the consolidated data points from prompt 1, Slot B on prompt 2, etc. The co-facilitator can start with just copying all three small groups' data on prompt 1. Then the co-facilitator will need to consolidate all the repeats. This should create a manageable number of data points for each prompt. There are typically about 10, but there could be as many as 25 (depending on the number and inspiration of your small groups).

In Figure 9.9, Slots A–E are the consolidated SME prompts while the other tabs were the original small group worksheets. Based on the number of small groups at this large Threatcasting, there can be quite a few tabs at the end. Figure 9.9 has been cut short for illustration purposes.

| ◄ ► | Prompt A | Prompt B | Prompt C | Prompt D | Yellow Pawn | Green Pawn | Blue Pawn | Black Pawn | Pink Pawn | + |

Figure 9.9: Visual of a combined RSW's tabs (post Research Synthesis).

Note: Walking Breaks

Build in breaks during the Threatcasting process. During the two-day workshop, participants are provided with places they can easily walk to that are filled with nature, interesting sites, or shopping. We do this because there is a lot of sitting and writing during a workshop, and people still need to move. We call these walking breaks and use them so that participants don't get stagnant during the exercises. We encourage participants to take a walk.

9.4 PHASES 2 AND 3

Futurecasting and Backcasting

See Chapter 4 in the methodology section of the textbook for a refresher on the EBM that the participants will now construct.

Using the combined RSW, participants once again convene in their small groups to start the next phases of the workshop. Generally, the facilitator will walk the participants again through the process they are about to partake in and step them through the sections of the TCW. It is also important to have technical assistance standing by in case participants have issues accessing the new workbooks.

As a point of emphasis, we use a pale-yellow background color for all the cells we expect participants to fill in on the workbook. This helps guide them to placing the information in the correct locations, which makes the data hygiene step in the post-analysis phase significantly easier.

9.4.1 CAPTURE THE RAW DATA

Picking the Foundational Data Points for the Workbooks

For many participants, picking the data points from the RSW for the EBM can be cumbersome and time consuming. Time is short when the participants are modeling their futures, therefore it's important to help them pick as quickly as possible.

Additionally, the goal is to have the participants use data points that could be challenging. The analyst(s) don't want the participants to cherry pick the data points they most prefer or know something about. It is ideal when the data points challenge the participants to consider and discuss how each of the data points interact with each other. Often, analyst(s) get the best results when the data points the participants use conflict with each other. Resolving that conflict yields valuable raw data.

Finally, a level of "randomness" is ideal for the participants to drive their EBM. The level of randomness depends on the method or tool provided to the participants. The method can be truly random or at the very least give the appearance and feeling of mild randomness.

The following are tools and methods that can be used by the facilitator and analyst(s). Process Note: For the following tools to work, each of the data points in the final RSW needs to be assigned a number.

Random Number Generator

This is a truly random approach for the participants to choose the data points. A random number generator can be found for free via an internet search. Typically, the participant enters the number range that corresponds to the number of data points in a specific section of the RSW. When the generator gives them a number from the range, the participants then use the data point next to the corresponding number.

This is repeated for the number of data points needed (i.e., the number of prompt categories).

Rolling Dice

Rolling a multi-sided die is one of the more popular ways for participants to choose their data points. It is also an interactive and fun way for the participants to get to know each other and for them to interact with the facilitator.

The only difficult part for using dice is that it is necessary to supply each small group with a die that has as many numbers as the number of data points in the RSW. Sometimes this can be quite large, so be prepared.

Facilitators can work around this by using a 10-sided die and having the participants roll the die multiple times. Each roll represents a single number. For example, if a participant rolls a 1 on the first roll and a 5 on the second, they would use the data point that is numbered 15.

Drawing Playing Cards

Similar to rolling dice, drawing cards is a fun interactive way for the participants to pick their data points. Just like using the 10-sided die, the participants use playing cards numbering 1–10. The participants then draw a card for each digit. If the participant draws the 2 of clubs and then the 6 of clubs, they would use data point 26.

It's Not Science or Math—It's slightly random and mostly fun

It's important to remember that the latter two methods are not intended to be scientifically or mathematically random. The goal is to first quickly get the data points picked and second to give the participants a reason to interact with each other and the facilitator.

Building the EBM with "A person, in a place, experiencing a threat"

Getting started is always the hardest part for the participants, especially if it is their first time filling out an EBM. Often the groups want to pick the "right" person or get their threat future perfect before they begin.

The facilitator should encourage them to just "jump right in." Giving their person a name generally helps. Having the group pick a specific city also helps get them started and also narrow down the expanse of ideas. Participants can feel overwhelmed with the blank page so getting something written as soon as possible helps to break the ice.

As we explored in Chapter 4, the questions are meant to encourage the participants to tell the best story they can tell. In this way, the secondary/sensing questions can give more detail (e.g., What will it smell like when the person experiences the threat?)

Tailoring these questions right from the start, based on the Threatcasting Foundation, can expand the quality of the raw data in the EBM.

Stories from the Lab

A Different Approach to the EBM

By now you know the basis of the EBM is a person, in a place, experiencing a threat. Moving from the high-level research prompts to a very specific future is often freeing for participants. They do not feel the pressure of identifying the single most important possible or potential future. That is why the facilitator will typically push each small group to focus on the person, in a place, experiencing the threat.

But what if your person isn't a person?

One of the strengths of the Threatcasting method is that it can also serve as a framework (see Chapter 1), meaning that the analyst(s) can adjust some of the participant instructions depending on the Threatcasting Foundation or even feedback gathered from the group during the workshop.

It was this kind of feedback that put a twist on that person in a place experiencing the threat.

On the second day of a workshop, a small group of participants called the facilitator over and asked that very question: "What if your person isn't a person?"

The group was exploring the future of weaponized artificial intelligence. They wanted to know if the AI could be the "person." What if the autonomous AI (the person) on a specific computation system (a place) was experiencing the threat? The group argued that because the AI was autonomous it would experience a threat alone much like a human being would experience it.

They were not trying to say that the AI was the equivalent of a person, but from an EBM stand-point it seemed like it could work.

The facilitator thought it was an interesting idea and encouraged the group to give it a try. The results were interesting, and the EBM served as an effective template to explore the nature and ramifications of the threat.

Since that workshop, facilitators have had participant(s) identify a wide range of persons. These have included:

• entire cities as a collective "person;"

• planet earth could be a "person" experiencing the threat of climate change but specific to a coastal region or desert ;

• robots;

• autonomous vehicles;

• drones;

• and many more

It was another interesting evolution of the Threatcasting Method, allowing the participants to use the EBM to really expand their imaginations and explore an even wider range of possible and potential threats.

Experience Questions

Typically for a large group Threatcasting workshop, the analyst(s) will have crafted the experience questions in Phase 0 too broad. Because the group is so large, the diversity and experiences of the participants will be wide and vary greatly. These more open-ended experience questions allow for the groups to capture the various perspectives of the participants. Don't forget, you can always tweak these questions in successive iterations of Phases 2 and 3 during the workshop if they don't initially gather data to help answer the Threatcasting Foundation.

However, just because the questions can be broad, that doesn't mean the participants shouldn't be as specific as possible. This can be a challenge for the small working groups. It is up to the facilitator to encourage and remind the participants to use their expertise but also to be as detailed and specific as possible.

Enabling Questions

Like the experience questions, the enabling questions for a large group workshop can also be broad. However, the aim for this approach is different from the experience questions. The enabling questions in the EBM are used to specify what has happened or what is needed to bring the threat about. Rooted in EBOs (see Phase 0), the domain expertise of the participants and the curation of the small groups takes on added importance.

Again, the facilitator needs to remind the participants that their professional backgrounds here do matter. The more specific they can get, pulling from real work experience, the more useful the raw data set will be for the analyst(s).

Often, asking clarifying questions can help.

- Is there a specific law or regulation that was or was not passed?

- What group, industry, or market might need to develop a specific technology?

- Are there specific areas of investment that needs to happen from specific organizations or countries?

- Have you seen something like this before but applied to a different threat or future?

Usually the participant(s) know more and have more detail in their minds than they are entering into the EBM. A few simple questions can help spark them to explain more.

Gates, Flags, and Milestones

Backcasting is the final part of the EBM. This is the area where most small groups get bogged down, tripped up or struggle to finish. First, this is usually a casualty of their time management. The groups will spend more time on the opening sections of the EBM and not save enough time to explore the gates, flags, and milestones. By the second day of the workshop (if there is one) the groups will have usually gotten better at time management.

The facilitator should monitor the progress of each group, encouraging them to save time for the final section. It can also be helpful to suggest the group does not have to "go in order." Meaning, the small group can skip around the EBM, filling in different details and information that strikes them. Giving them the option to race through the EBM as quickly as they can to get to the end, followed by the option to revise, edit, and add has been an effective strategy to get more detail and more robust gates, flags, and milestones.

The facilitator should also urge the participants to "name names." Meaning, as the group explores the gates and milestones, they should consider who is going to take this action, where they are and what their purpose is.

Sometimes being vague is alright. Even if the participants don't have the specific expertise or understanding for how an action could and should be taken, they should write down as much information as they can. This can be a call out that more work needs to be done.

Example vague language that can be effective:

- a broader cultural conversation must be started to explore…

- academia needs to research the possibilities or perils of…

- an industry consortium needs to be convened to define the problem of …

These vague approaches and suggestions can give specific organizations (e.g., industry, academia, etc.) a place to start. They also allow the analyst(s) to call out the needed actions in the post analysis or follow up with SMEs to get more detailed steps to be taken.

Stories from the Lab

Running with Microphones

When the small groups come back together in the larger workshop, the report outs can be an opportune time not only to share what each group has been discussing but it can also be a chance to have some fun, bringing the group together.

Limiting each small group's report out to just a few minutes has a primary practical usage when conducting a large group workshop. By limiting the time for each presentation, the facilitator can keep the group on schedule while at the same time giving all the participants a window into the various possible and potential threats that the collective participants have been discussing.

Also, to facilitate this time in the workshop, we often use a microphone. The microphone also has two pragmatic purposes. First, it allows the participant who is reporting out to be heard in a large room filled with chatty people. Second, the facilitator can also take the microphone away when the presenter's time has expired, allowing the room to stay on schedule.

Often by the second day of a large group workshop the participants start to have fun with their report out. They are more polished. Sometimes they will work together, employ visuals, and even stage a small one-act play. And this is all done under the time pressure of the clock. The collective pressure and game lets the participants root for each other and keeps the mood light and energized.

However, there was a large group many years ago that took an unexpected twist. This was a particularly large group, over 100 participants. There were multiple facilitators. But true to the process there was only one microphone. When one presenter was wrapping up and getting close to the end of her time she still had more to say. The facilitator slowly walked toward her with their hand out to take the microphone. The timer was beeping. The facilitator reached for the microphone and suddenly the presenter took off running. She continued to present the work her small group wrote, but did it jogging around the room. The facilitator, who was a runner, chased the presenter while the entire room erupted into laughter.

When the presenter was finished she stopped running and handed the microphone back to the facilitator. With a smile she said, "Sorry. It was just a really good future and I wanted to get it all out."

Report Out

Just like the report outs at the end of Phase 1 (Research Synthesis), each small group is given the main stage to share their EBM (vision of the future). Using a timer (perhaps 90–120 seconds) ensures that everyone has a chance to share in a concise manner. It is also a camaraderie building activity as they race to share their story about their person before the timer goes off.

Given that they generally share more details than might be written down, it might be helpful to have the co-facilitator taking high level notes with an eye toward the post analysis. This is also a good point to remind the participants to provide more details in their written product to reflect all the good ideas that they have come up with during this activity.

9.4.2   RINSE AND REPEAT

Typically, this process (futurecast and backcast) is repeated two to three times during the course of a Threatcasting workshop. Typically, the Day 1 schedule provides enough time to get through one iteration of a futurecast and backcast. As the participants will be better acquainted with the process on Day 2, generally you can execute another two iterations of futurecast and backcast. It is important to note that for each of these EBM developments (i.e., rinse and repeat of Phase 2 and Phase 3 actions), the small groups must select new data points from the prompts. This will result in new and unique EBMs for each round. The more rounds that can be accomplished, the greater the volume of raw data for the post analysis.

Normally, after the first iteration of the process, the core team will come together to determine if any modifications need to be made to the workbook questions for the remaining rounds to ensure data collection is in alignment with the Threatcasting Foundation. Typically, this might be an added constraint on location or a person's job or even changing some of the enabling or experience questions.

Ultimately, we generally make a couple of modifications for each round to keep the process lively and to challenge the participants as they get more and more comfortable with the Method.

For each round of the EBM construction (futurecast and resulting backcast), you should create a new workbook to be shared with the participants.

Timing

As mentioned earlier in this chapter, you should aim to have participants complete an EBM (both futurecast and backcast) on Day 1. Depending on your schedule, perhaps allocate an hour for the small group work and then 30 minutes for the report outs. The main intent is to get participants comfortable with the workbooks and process as soon as possible. This should leave you time at the end of Day 1 for wrap up and review of the next day. Then, on Day 2, you should be able to schedule longer time periods for rounds of Phase 2 and 3.

On Day 2 a typical schedule might look like the following:

8:30 a.m. to 9:00 a.m.	Gather and Coffee
9:00 a.m. to 9:30 a.m.	Kick Off, Agenda Review, Review Previous Day
9:30 a.m. to 11:00 a.m.	EBM Creation (Futurecast and Backcast)
11:00 a.m. to 11:45 a.m.	Report Out
11:45 a.m. to 12:15 p.m.	BREAK and Lunch Buffet
12:15 p.m. to 2:00 p.m.	EBM Creation (Futurecast and Backcast)
2:00 p.m. to 2:45 p.m.	Report Out
2:30 p.m. to 2:45 p.m.	BREAK
2:45 p.m. to 3:30 p.m.	Reflections, Wrap up, and Next Steps

The Shifting Role of the Facilitators

The facilitator's role during the beginning of the workshop (Day 1) is to be an explainer and resource. Many participants will not have experienced a workshop, so they will have many questions. Over communicating, patience, and a little good humor is a good strategy. Sometimes good ideas and interesting perspectives can come from these early discussions.

As the workshop continues (end of Day 1) and the participants get more comfortable, it is important for the facilitator to keep the workshop to schedule, encouraging the participants to be open, use their imaginations and consider the perspectives of their fellow participants.

The final role of the facilitator (Day 2) is to monitor the progress of the small group's development of the EBMs, jumping in to encourage them to add more detail, professional insights, and leave time for all sections of the EBM.

At the close of the workshop, over communication about what the participants can expect next will help answer many questions now that they have given so much time and energy to the workshop. Ending the workshop with a discussion, reflection, and gratitude will honor the participants' contributions and also encourage them to continue working with the analyst(s) during the post analysts and peer review phases.

9.4.3   CLOSING THE THREATCASTING WORKSHOP

After the "Big Tent" workshop concludes, there are a few immediate actions to take. These will be for the team with the venue event staff. You want to get invited back, so it is important to tidy up and follow the venue's rules.

Action

Build a checklist to walk through the closeout.

1. The room will need to be returned to the way it was found (tables, chairs, screens in their original positions).

2. All trash cleaned up.

3. Catering removed and disposed of.

4. Charging cords removed from sockets and appropriately returned.

5. Hold a closeout session with the venue staff to ensure paperwork is signed and closeout procedures are met. Thank them!

Bonus: we often give the venue staff a thank you gift of chocolate or seasonal baked goods. Extra touches like this are our commitment at the lab to always be humans first.

Additionally, you should consider doing a quick "After Actions Review" (AAR) with the facilitator(s) and analyst(s) that supported the workshop. Yes, you will be quite tired after running the event for the last two days and you will want to spend some time with the participants before they depart the local area. However, building in a short time with the staff to get initial feedback and assessments is critical as many of these comments/ideas will be forgotten if you wait too long. Additionally, we like to take the opportunity to confirm next steps/timeline for the start of Phase 4 activities.

9.5   PHASE 4

After completion of the Threatcasting Workshop and the generation of the raw data, the next phase in Threatcasting is post analysis.

Identify Advocates

A large group Threatcasting Workshop provides participants with a new and exciting method to think about the future to take back to their organizations. The participants are encouraged to implement the findings within their organizations. Workshop participants frequently leave the large group Threatcasting Workshop events with defined research areas they are interested in pursuing further, thus creating champions of the Threatcasting process who then become advocates of the process.

9.5.1   ANALYST(S)

Because of the volume of raw data developed in the large group Threatcasting Workshop, it can be helpful to have multiple analysts perform the post analysis.

It can also be helpful to have the analysts attend the workshop so they are familiar with the report outs and the people who are contributing. The ideal analyst is one who has diversity of thought, knows about many different things, and has the ability to see from different lenses and angles.

Action

Build a Timeline.

The analyst(s) will need to craft a timeline for the post analysis that supports the required completion date of the final products.

1. Coordinate with all analysts on the project on their availability.

2. Determine whether to conduct the post analysis in-person or remote.

3. Schedule a first review session to take place 1–2 days following the workshop.

4. Provide ample time for the Analysts to conduct their individual analysis of the raw data.

5. Convene analysts together for collaborative sessions periodically through the process.

6. Draft findings.

7. Establish reviews of the findings with the core team and participants.

8. Finalize findings.

9. Present findings to steering committee.

10. Begin work on final output.

Post-Analysis Workbook

The post-analysis workbook is a part of the project documentation, typically a spreadsheet. It is a place for the analyst(s) to capture each round of the post analysis (summary, meaning, novelty). It is beneficial to use this so that it can be shared with other analysts and also so that it can be included in the final output, assuring transparency in the methodology. Therefore, just like the other workbooks used throughout the Threatcasting Method, we use Google Sheets for the post-analysis workbook also.

As seen in Figure 9.10, this is the simple template for what your post-analysis workbook might look like.

	A	B	C	D	E	F
1	Round	Team	Summary	initial themes	meaning / insight	novelty
2	1	Blue				
3	1	Pink				
4	1	Orange				
5	1	Yellow				
6	2	Yellow				
7	2	Blue				
8	2	Orange				
9	2	Pink				

Figure 9.10: Example of a post-analysis workbook.

Kick-off event for the Post Analysis

Just as you did for the Threatcasting Workshop, you should start the post-analysis phase with a kick-off meeting. Here you can present the timeline, locations of the raw data, locations of the SME prompts, and any other related information that the analysts might need during the course of their post-analysis activities. Even though the first part of the post analysis is done individually, it is important to come together as a group at the beginning.

Stories from the Lab

Two Days in Portland, OR

Depending on the Threatcasting Foundation and the number of analysts, the post analysis can be as interactive and lively as the workshop that developed the raw data sets.

In the lab we were working on a sponsored research project for the Army Cyber Institute exploring the future of weaponized artificial intelligence (Johnson et al., 2017). The application areas in the Threatcasting Foundation were broad, stretching between military, government, industry, trade associations, and academia. Because of this we had a range of analysts who all participated in the post-analysis phase.

To gather everyone's input and perspectives we had each analyst conduct all the steps of the post analysis phase individually. Each of us prepared to report out our findings to the rest of the analysts and then work collectively to generate our findings and the outline for the report.

The amount of raw data and post analysis was large. To provide the time to focus on exploring each analysts' work, the lab curated a two-day post-analysis event in Portland, OR.

Over two intensive days:

- *Each analyst presented their post analysis and proposed findings (Day 1).*

  - *During this portion of the event, it became clear that each analysts had not only a different perspective but also different tools and processes for conducting their post analysis.*

  - *These different approaches and perspectives expanded the impact of the final report.*

- *We clustered the findings and discussed the implications (Day 1).*

  - *Here we reflected on our different approaches and how that might affect the clustering.*

  - *We argued, disagreed, agreed, came to consensus, and then disagreed some more.*

- *We had a great dinner together (End of Day 1).*

- *We broke into smaller teams to revisit the clusters (Day 2).*

  - *Each analyst self-selected smaller teams.*

  - *Typically, the analysts that shares similar approaches worked together while others decided to work by themselves*

- *We developed an opinion and initial findings (Day 2).*

  - *Ultimately the lab needed to come to an agreement, even if we didn't all agree.*

  - *Where we didn't agree it was noted in the report.*

- *We began the outline for the final report (End of Day 2).*

  - *Using the findings, we began to draft the final report, assigning each analysts different sections to begin writing.*

Although our two days in Portland were highly productive and scheduled, we also allowed time for collaboration, disagreements, and discussions. For some analysts the post-analysis phase can be broad and open ended. But with an expansive Threatcasting Foundation and a high volume of raw data sets it can be helpful to structure the post analysis as tightly as one would schedule the workshop.

9.5.2   PREPARE THE RAW DATA

Data

The positive of hosting a large group Threatcasting Workshop is it generates a large dataset. The output is quite lengthy and dense because of the dataset's size. The dataset is so large that the management, protection, and storage become extremely important. Having a plan for this at the beginning of the Threatcasting Workshop journey is important.

Data Hygiene

Assign one team member to conduct the data hygiene on all the data in order to prepare it for the post analysis. Chapter 6 in the methodology section of the textbook can provide a refresher on these steps.

Action

1. Immediately following the completion of the workshop, ensure that you have a complete copy of all the RSWs and EBMs saved in an off-line (but protected) location. This will ensure that if data is accidentally corrupted during Phase 4, it can be recovered.

2. Write down your data management and storage strategy and share with the analyst(s).

3. Once the assigned team member conducts the data hygiene step, create a complete copy of all the RSWs and EBMs and save in an off-line (but protected) location.

9.5.3   PERFORM THE ANALYSIS

Post Analysis

Typically, each analyst will work alone on the three steps of post analysis: summary, meaning, and novelty. Once the three steps have been captured, the analysts convene and present the results to facilitate a group analysis on the raw data. Generally, this can be done over the span of two to three days (depending on availability of the analysts).

Every analyst has specific tools they prefer to use and a preferred environment to work in. However, it is important to capture their analysis.

Action

1. If there is only a single analyst: Review your notes and any project documentation that would help give an additional perspective before you begin.

2. If there are multiple analyst(s):

   a. Document and discuss each person's expertise and how they are thinking about approaching the raw data.

b. Set a timeline for each analysts' solo post analysis.

c. Select a shared project documentation template so that each analyst is using the same format in the end.

3. Write down the Threatcasting Foundation on a sticky note or piece of paper so that you can refer to it throughout the process.

4. Periodically step away from the data to clear your head.

5. Remember to use the techniques (described in Chapter 6) for each round.

6. Keep notes in the project documentation for any insights, ideas, or early clusters for yourself and other analysts.

9.5.4   GENERATE, VALIDATE, AND REVIEW THE FINDINGS

Generate

Following the three steps of post analysis as described in Chapter 6, the analyst(s) will generate their findings. Using the Threatcasting Foundation as a guide, the analyst(s) reviews Round 3 "novelty" and considers how to answer the Research Question as it is applied to the Application Areas.

For a large group workshop there is generally a large amount of data and a high number of people who will review the findings. It is important to remember that it is the lead analyst's role to have an opinion based upon the initial research—SME interviews combined with the raw data sets from the workshop.

Validate and Review

To validate the findings the analyst(s) first explores if the findings answer the research question and application areas. Next, the analyst(s) returns to the initial research and SME interviews to validate the findings and also identify if there are any outliers or missing perspectives.

Integral to the validation and review of the findings for a large group workshop is communication with the core team, steering committee, participants, and SMEs. Because the number of people and perspectives are so large, the analyst(s) should gather as much feedback and comments as possible.

9.5.5   COMMUNICATION

Core Team and Steering Committee

It is important to keep the doors of communication open between the analyst(s) and the core team/steering committee during the post-analysis. Confirm with the various core team members how they wish to be involved in the post analysis. Some core team members are curious and will appreciate updates throughout Phase 4 and others prefer to see the draft results at the end of Phase 4. Also, let the steering committee know when you expect to have some raw results of the post analysis for their feedback so that you can schedule an update meeting. Ultimately, you might ask these groups to conduct a peer review of the initial findings and to also sign off on the final report before it goes public.

Participants

It is also important to maintain communications with the participants during the post-analysis phase. First, we want to give them the opportunity to share any closing thoughts about the venue, process, and event. Next, we need to ask them how they want to be attributed on the final products. Some participants are fine with listing their name and organizational affiliation, while others prefer to remain anonymous. Many participants will have multiple affiliations, so this is the opportunity to determine which one they want to use. Finally, the participants should be given an opportunity to serve as peer reviewers on the findings.

9.6   PHASE 5

Over the years, we have used combinations of all the Threatcasting output types (as discussed in Chapter 7) in large group Threatcasting events. Partially this is due to the sponsor audience needs, but also because these larger sized events tend to have a larger budget associated with them. Recall that your Threatcasting Foundation should help define what the final products of your Threatcasting should be.

Action

1. Review your Threatcasting Foundation to determine how your sponsor wants to receive the findings.

9.6.1 TRADITIONAL OUTPUTS

Stories from the Lab

Many of our large group Threatcasting Workshops end with a final report. Over the years, we have increased the design elements on the report. We now work collaboratively with a graphic designer the minute the text and data are ready. We meet with the designer and go over what the research question was, what some of the SMEs covered, a few of the threat futures, and the executive summary. We then hand the report over to the designer to read any and all other parts. We work back and forth collaboratively to achieve an outcome that is on brand for the Threatcasting Lab and unique to each sponsored report. The designer custom creates the cover and interior for each report.

For each report we include photos of the spaces that we used for the workshop, images from the city or location, and anonymous pictures of the participants collaborating within the design of the report. Therefore, generally one of the analysts or facilitators will bring a nice camera to the workshop to take some "artistic" photos of the surroundings. We don't take any recognizable photos of the participants as some will prefer to be attributed as anonymous in the final report. Examples of these design decisions can be found on the ASU Threatcasting Lab website in our final technical reports.

Technical Report

The most typical output of a large group Threatcasting is a technical report. Section 7.1.2 has more details on content and scope. A professionally written report is a typical deliverable to show the value the sponsor organization received for the cost of the event. Understanding the Threatcasting Foundation and the nuances of the domain space will also help you to incorporate design graphics to enhance the report.

It can also be helpful to use a tone/copy editor once you complete the writing. If there are multiple analysts and authors combined with feedback from participants and SMEs, an editorial pass will help smooth out the language. The person who does this pass can be a professional or trained editor. But they can also be a person "with fresh eyes" who has not been involved with the project.

Academic Papers

For a subset of the findings from the large group Threatcasting, you might want to write an academic paper—especially if one of the gates is the requirement for more academic research on a specific topic. Publishing a paper or short article in an academic journal within that discipline could be very helpful for disseminating your findings.

Because of the large number of people involved in the large group Threatcasting, there will be more opportunities for analyst(s) to partner with people to write academic papers. It can be useful to start this process early. During the workshop or during post-analysis activities, if an analyst comes across a novel idea or finding that is shared with a workshop participant or SME, they can reach out to see if that person would want to work on a paper together. For many new analysts and students this is a good way to get started publishing in your academic career.

Action

1. During the SME interviews, workshop, and post-analysis activities, write down novel ideas and findings that might work as the foundation for an academic paper.

2. For each idea, brainstorm different participants and SMEs who might be good collaborators for the academic paper.

3. For each idea, brainstorm a list of possible publications that might be appropriate for the paper. Record the following information on these publications:

   a. Publication Name

   b. Website

   c. Submission Deadlines

   d. Submission Criteria (including length, format, etc.)

   e. Any associated costs

4. Get to writing!

Briefings, BLUFs, Executive Summaries

A BLUF (Bottom Line Up Front), a summary, or a briefing offers a window into the high level of the analyst(s)' findings. Because the large group Threatcasting workshop has so many people on the core team, steering committee, SME interviews, as well as the participants, it means that there will be many different opportunities for the analyst(s) to present and report out the findings.

As the analyst(s) finalize the findings and the report, they should consider offering a briefing to any person or organization that has been involved in the project. Aside from being a professional courtesy for participating, it is also an excellent way to socialize the findings and increase the social network for the analyst(s).

Refer back to Section 7.1.2 for more details.

9.6.2   ALTERNATIVE OUTPUTS

Graphic Novellas

A graphic novel is a story presented in comic-strip format and published as a book. A graphic novella is a short version of a graphic novel. Those who do not have the time to read the entire technical report, nor a 20-page academic paper, will likely have the time to read a 10-page graphic novella. Typically, this will be enough to whet their interest in starting conversations about the topic within their community or organization or reading the other papers. Section 7.1.3 has an example of a graphic novella generated from Threatcasting results.

These graphic novellas can be highly impactful because they can show the reader what a possible or potential threat future could look like. Because it is a visual medium, the writer and creators can pack in a lot of information that was developed in the workshop and written about in the technical report.

For more information about the construction of graphic novellas, see *Dark Hammer: A Retrospective of Science Fiction Prototyping* (Army Cyber Institute, 2018).

Developing a graphic novella takes a different skill set than writing the technical report and academic papers. The lead analyst might be able to find a production team (see below), or if there is a sponsor for the Threatcasting workshop some budget might cover hiring the team.

Action

1. Review the technical report and raw data set to determine if it would be impactful to develop one or more of the threat futures into a graphic novella.

2. Determine if you have access to a team who can help develop the graphic novel. If you are in an academic setting you might be able to find a team to work with in the arts or design departments. If you are in a corporation or organization there may be a team in the marketing or advertising department.

3. If you do not have access to a team, determine if you have the budget or funding to hire a production team.

The production team for a graphic novella will vary depending on the project as well as the talent and skills of the team. The following skills could be found in a single person or a larger production team.

- Writer—the writer uses the Threatcasting Technical Report and raw data sets to write a fiction script that captures the threat futures.

- Artist—the artist uses the script the draw the panels.

- Creative Director—the creative director works with the artists to develop the look and tone of the story.

- Executive Producer—if the team is large, a producer can help keep everyone on track and on schedule.

- Sketch Artist—a sketch artist can be used at the very beginning of the process to draw quick visualizations.

- Pencil Artist—the pencil artist takes the sketches and refines them in more specific black and white drawings.

- Ink Artist—the ink artist refines the drawings to near final version.

- Colorist—The colorist adds color and shading to the inks.

- Lettering—The final phase is to add the lettering or dialogue to the graphic novella.

Action

1. Project Kickoff: Conduct an initial scoping meeting with the design team. Determine their typical process, the project milestones, anticipated deliverables, and the required input you will need to provide.

2. The writer uses the technical report and raw data sets to craft a graphic novella script.

3. Once reviewed and approved by the rest of the team, the sketch artist can draw the initial rough set of visuals according to the script.

4. The visuals then go through multiple rounds of revisions and refinements (sketches, pencils, and inks).

5. When the visuals are locked and complete the colorist adds color and shading.

6. The final step is to place the dialogue, narration, and descriptions.

7. Many of the graphic novellas from the lab include an Introduction and/or Afterward to position the story and how it relates to the technical report.

Podcasts and Interviews

The Threatcasting findings can be presented in audio form. This medium appeals to a large audience and has been growing in popularity. Many individuals might not have time to read reports or view graphic novellas during their workday, but they often have time to listen to podcasts in their workday. It is also a good way to draw attention to your research and ultimately refer audiences back to the technical report if they want more information.

There are two ways to approach podcasts and interviews. The first would be to find existing podcasts that might be interested in interviewing you about your findings.

Action

1. Survey your core team and steering committee to determine what podcasts that they listen to and which their boss(s) listen to.

2. Research these offerings to determine if a component of the findings would make an interesting episode.

3. Contact the podcast creator with a short introduction that refers to your report.

4. Practice, practice, practice what you are going to say before the interview. Develop a couple "go to" phrases that you want to work into the content.

5. Make sure to reference your technical report and provide the podcast creator with links for where the audience can access the report.

6. Keep a record of your interviews and use them to contact other podcasts, using the previous interviews as examples of your work.

The second approach would be to record your own podcast and explore the findings in the technical report. You can record this alone or include the additional analysts, SMEs, or participants.

Action

1. Gather the necessary technical equipment to record the podcast (e.g., computer, podcast app or software, microphone).

2. Listen to similar podcasts to get a feel for length and tone that will appropriate for your podcast.

3. Write a script that outlines the various points you would like to make from the technical report.

4. Invite quests (e.g., core team, SMEs, participants) to join you and engage in a discussion about the findings.

5. Post the podcast on as many sites and services as you can to give people the most access.

6.   Follow up with your guests and have them share the podcast with others.

7.   Often people learn that a single podcast is not enough to capture all of the findings. Consider doing several "episodes."

Final Note

The most important thing to know about hosting a large group Threatcasting Workshop is to communicate with everyone involved often; it will make the entire experience a better one for all involved.

CHAPTER 10

Small Group Threatcasting Workshop

"I do my best because I'm counting on you counting on me."
—Maya Angelou

10.1 INTRODUCTION

A small group Threatcasting Workshop varies from the large group workshop first and foremost because of the number of participants. Small group Threatcasting Workshops can be used to more deeply explore threats and research a large group workshop identifies. Smaller organizations and collections of people who need to rapidly identify a range of possible threats also use the small group Threatcasting Workshop. Typically, a small group is used on specific projects where threatcasting is applied to a single threat, product, problem set, or company. A small group allows the participants to remain somewhat nimble, adjusting on the fly quickly.

The disadvantages of a small group are limited perspectives. When a limited number of practitioners are in the room, the data output is less than that of a larger group. That said, deep and repeated collaboration across all of the participants will build a more robust set of data. The goal of the small group workshop is to bring a small group of diverse multidisciplinary participants together. These practitioners look at a broad intersection of factors and potential threats to develop a collection of threat futures as expansive as possible.

The coordination and curation of the room participants take less time than a large group Threatcasting Workshop because it is easier to coordinate five to twelve people's schedules.

Those participating in small groups threatcasting exercises are typically accustomed to looking at specific problems that explore a new idea or provide better clarity on a previously identified threat. When constructing small group sprints, curation of the people in the room is crucial as participants from diverse backgrounds and domain expertise are vital to capturing the most comprehensive data generation with the widest variety possible in potential futures.

Small groups are typically 5–12 participants, although they can be curated to include a few additional participants. With a smaller group, several experienced "threatcasters" and an experienced facilitator are helpful. Choosing who is a part of a small group workshop can be even more important than the curation with the large group workshop.

Physical vs. Digital

The most significant advantage of a small group is the limited number of participants, making the Threatcasting Workshop easier to coordinate and quicker to conduct. Because the size is smaller, a small group Threatcasting Workshop could be held either physically or digitally. Both versions have their advantages and disadvantages.

In-person threatcasting allows individuals to be completely mentally focused and also aware of the physical cues from other participants. The small group work can also be accomplished rapidly in one or two days. The digital version of the workshop typically takes longer to conduct, as it can be difficult for people to spend 8 hours straight in a digital or virtual environment. Therefore, it might be scheduled as 1- to 2-hour blocks over 1–2 weeks. However, the advantage is that a more diverse set of participants can be included. If the work is spread out over a week, with hour-long working sessions each day, then the participants have more time to populate the EBM with details. Performing a small group threatcasting as a digital event also means that some of the work can be done asynchronously if needed.

Stories from the Lab

Up until this point in the book we have mainly discussed physical Threatcasting Workshops, events where people gather together in a room for a set period of time to model threats. However, not all Threatcasting Workshops need to happen in person.

Some small group Threatcasting Workshops can function as digital or virtual events. There are some key changes or modifications to consider when hosting a digital small group workshop.

- *You will need a digital or virtual meeting space (e.g., Zoom, Microsoft Teams, Apple Face-Time, etc.)*

- *Spread out the workshop over 3–4 days with 2-hour working sessions each day*

- *Prioritize collaboration time.*

  - *Distribute videos and pre-work before the event.*

  - *Consider recording videos to give participants instructions so that they can watch on their own time.*

- *Try your best to set expectations with the participants for how many hours you are asking of their time. Getting them committed to do the work.*

There are pros and cons to this approach. The analyst(s) should consider the following.

Pros

- *The participants don't have to be in the same space geographically.*

  ○ *Participants do not need to travel, saving time and money.*

  ○ *Analyst(s) can pull from a wider range of people.*

- *Participants can take more time to work in small groups when they are not bound by a specific amount of time in a room together.*

- *The conversations that happen in the virtual breakout rooms can be different, more honest or in depth because people are not physically in the same space.*

- *Some participants prefer not to be in person.*

Cons:

- *The facilitator will need to work harder to monitor and encourage participants.*

- *Coordinating participants' calendar over a longer period of time can be challenging.*

- *Because the participants are not in the same room, they can be distracted by personal and professional interrupts.*

- *Some participants prefer to be in person.*

10.2 PHASE 0

Preparation and Curation

The Threatcasting process is built by a series of actions, each building upon each other. To complete an entire Threatcasting from beginning to end requires a plan. Generally, the plan for Phase 0 will be very similar to what is required to prepare for a large group Threatcasting Workshop, however some of the logistics are easier.

10.2.1 DEVELOP THE THREATCASTING FOUNDATION

For a small group Threatcasting Workshop, the specific definition of the Threatcasting Foundation will guide the analyst on each step and decision. No matter the implementation size of a Threatcasting Workshop, the development of the Foundation remains the same. See Sections 2.1.1 and 9.2.1 for additional details.

Pick a Topic

The problem or threat you convened the small group to explore becomes the topic of the workshop. Having your topic in place helps to select the language used for the invitation, the workbooks, and images to be used. The topic also determines the application of the methodology and the desired outcome.

Action

Refresh yourself on defining the topic by visiting Section 2.1.1 of the methodology section of this book.

Research Question

Once the topic area is selected, the research question aims to narrow the focus to something that can be successfully explored using a small group Threatcasting Workshop. If the research question is too broad, unclear, or not specific enough, the Threatcasting Method will not yield the optimal results.

For a small group Threatcasting Workshop, the research topic is generally more focused than in a large group. Whereas a large group question is broad because there is a broader range of people, a small group research question is generally more focused and tailored to the smaller group of participants. This focus will allow the small group participants to explore the range of possible and potential threats with a greater depth. Because the analyst(s) will have fewer EBMs (i.e., raw data sets to work from) this more focused research question assures that enough data and perspectives will be developed in the small group EBMs.

There is a simple set of questions that can be used to help ground the research question.

Start by asking "Exploration" questions and open-ended "how" and "why" questions about the general topic. Consider the "so what" of the topic. Why does this topic matter? Why should it matter to others? Identify one or two intersecting questions that are engaging and could be explored further through research.

Next, determine and evaluate the research question. What aspect of the more general topic will analyst(s) and participants will explore? Is the research question clear? Is the research question focused? A research question must be specific but also leave room for complexity. Questions should never have simple yes/no answers and should require research and analysis.

Action

For examples of good and bad questions, revisit Section 2.1.1 and Exercise 2.1.

Application Areas

The output of Threatcasting identifies not only actions that can be taken by the sponsor and other parties but also events, technologies, and changes that could happen over the next decade that will indicate whether we are moving toward or away from the potential threats occurring. Now is the time to determine what the data collected during the workshop will be used for. Will the raw data be turned over, will there be a technical report, will there be articles, podcasts, or sci-fi prototypes? Knowing this now will help build the foundation.

STOP HERE

Don't move forward until you can clearly write your research question and what the data will be used to create.

Action

Write each of these down to start your project documentation.

- Topic

- Research question

- Application areas

Review the topic and research question to see if they will inform the application areas. If needed, refine each of the three areas to be as concise as possible.

10.2.2   ASSEMBLE THE TEAMS

Core Team

The core team for a small group Threatcasting Workshop can look similar to a large group in the types/context of individuals that should be invited to join. It should include the Threatcasting analyst(s) as well as a representative from the audience or users in the Threatcasting Foundation's application areas. If the Threatcasting Method is being used for a corporate, industrial, military, or governmental use, the core team could include key team members who will use and/or champion the findings. Unlike a large group Threatcasting, there is often not a need for a support staff member to be part of the core team (due to the easier logistics and communications requirements).

A small group Threatcasting core team is typically between three to five people. If more people are added, team communications management can be too cumbersome for the analyst(s).

Each small group Threatcasting's core team makes all the decisions (including logistical decisions) and also signs off on the final output together. The duration of the entire project can take four to eight months. The shorter timeline (compared to the large group Threatcasting event) is generally due to the easier/shorter logistics and coordination timelines.

The core team will work through all the workshop details, including the topic selection, invitations, subject matter experts, participants, logistics, and deliverables. This team is also essential for serving as stewards and stopgaps for problems and early reports.

Finally, it is important to acknowledge and investigate the organization's culture to determine the best communication modes, history, and politics when building a core team. Knowing these details will ensure smooth-running meetings. What does this mean exactly? It means find out key things about the sponsoring organization. For example, there are many online tools that some organizations cannot use. Knowing what is usable and what is off-limits helps with confusion and extra work.

Action

Build your core team.

1. Create a list of who should be included on a core team. Do this by using titles, not names (e.g., sponsor decision-maker, Threatcasting coordinator, special advisors, sponsor coordinator). The core team are the people who will work closely with you throughout the project. Their inputs can help fill in the analyst(s)' experience or social network gaps.

2. Once you have the preferred structure of the core team, determine the individuals you want to invite to join the core team. Create a list of specific names to ask.

3. Create a formal invitation to send, including FAQs about what you are asking them to do on your core team. Sometimes a follow-up call will answer any questions they might have.

4. Determine the core team meeting frequency, such as bi-weekly and platform such as conference call, Zoom, Skype, Teams, etc.

5. Set a standardized time for each meeting. This will assist the core team members to protect that time on their schedules. Depending on the amount of time before the start of the workshop, meetings can be bi-weekly, but then will need to change to weekly as the event nears.

Steering Committee

A steering committee for a small group Threatcasting Workshop is not essential but can be helpful; it might only be one to two people. This group is in addition to the core team. The steering committee does not work on logistics, but instead supports the Threatcasting team by reaching out to their personal and professional networks. The steering committee brainstorms with the core team regarding the following questions.

- Is the Threatcasting Foundation sufficient?

- Who else should be on the core team and steering committee?

- What would be good prompts?

- Who would make good candidates for SMEs?

- Who would be good participants for the workshop?

- Where should the results be socialized?

The steering committee then reaches out to their vast networks to invite SMEs and participants. Steering committee members are chosen for each workshop uniquely. They work in a variety of domains and have a willingness to share their network with the Threatcasting team.

Action

Build your steering committee.

1. Creating a list of who should be included on a steering committee. Note, one way to start your list is to ask the core team to open their contact lists and offer to make introductions to the individuals most connected to the research topic.

2. Writing the email you will use to the connected individuals. The focus of the email is to ask if they would be willing to meet on a group call regularly to discuss the event.

3. Writing your first agenda for the first meeting of the monthly steering committee. Include things like discussing dates of the event, considerations for potential SMEs, and participants' names.

Additional Analyst(s)

The lead analyst is the person who will be in charge of the execution of the small group Threatcasting Workshop. They will work with the core team to set the Threatcasting Foundation and select the prompts for the workshop. The lead analyst doesn't have to be the facilitator, but often they fill that role as well. Most significantly, the lead analyst will run the post analysis and be the lead author on the output.

Early in the project planning, the lead analyst and core team should consider other additional analysts. These additional analysts can support the lead in all preparation tasks. However, the main role of the additional analyst is to provide a different perspective during the post analysis and output phases. Remember that the bias and expertise of the additional analysts is also a form of curation, providing points of collaboration and varying experiences.

Facilitator

Similar to a large group, you should select your facilitator early on in the process. Often, the lead analyst fills this role. Even though the number of participants is smaller, the facilitator role is still important to keep the workshop on schedule, explain the details of the method and workbooks, and encourage and prompt the participants. In fact, the facilitator for the small group Threatcasting Workshop will get to spend more time working with the participants because there are fewer people with which to interact. This means that the facilitator might change how they plan to prompt the participants, opting for a detailed discussion instead of a generally one-way briefing.

Project Documentation

Even with a small group and a simpler logistics/communication footprint, it is important to create a solid project documentation plan. Potentially either the sponsor or members of the core team might have restrictive IT processes that limit the available collaboration platforms or sharing services that can be used. Early in the project, create the data protection plan you intend to follow (i.e., backups, versioning, etc.). Further in this chapter we will discuss creating the various elements that will exist in this documentation.

Communication Plan

Building a communications plan is also essential to managing the planning of a small group Threatcasting Workshop. One could argue that it is more essential when working with a smaller group, but it also might be easier. This plan should outline how and how often information will be disseminated to the group (both with respect to the core team, the steering committee, and the participants) and who will be responsible for these communications. It helps set the general expectations and allows you to be a more effective communicator.

10.2.3　SELECT AND GATHER RESEARCH PROMPTS

Working to create the core team and steering committee, the analyst(s) begin to gather the appropriate research prompts to address the topic and answer the research question.

Prompts

Next comes the art of selecting prompts for the workshop. As discussed in Section 2.1.3, there are many categories of prompts such as:

- social science research

- technical research

- cultural history

- economic projections

- trends

- data with an opinion

How many prompts are ideal? The answer: enough to cover the problem space that you will be exploring. Don't artificially constrain yourself at the beginning of the planning effort. Instead, remain flexible—perhaps start with four and see how that evolves. In general, there are no differences in selecting the prompts for a small group versus a large group.

Action

- Visit Section 2.1.3 of this text's methodology section for a refresher on prompts.

- Determine one core team member who is in charge of research prompt collection; allow this team member to report in at bi-weekly core team calls.

- Conduct a brainstorming session with the core team to determine the buckets of prompts that could be useful to the Threatcasting Workshop.

Subject Matter Expert

For a small group Threatcasting Workshop, SMEs can be useful, but are not a requirement as a prompt type. The small number of participants and quicker coordination often doesn't leave time for the work that goes into gathering an SME. However, the depth and focus of an SME's inputs are always beneficial and should be sought if there is time. Often, if a small group workshop is convened following a large group workshop, so potentially some of the SMEs can be repurposed.

An SME is a person with particular expertise, perspective, or opinion that the core team, steering committee, and analyst(s) select to serve as a prompt for the Threatcasting Workshop. By interviewing an SME directly about the topic and research question, the analyst(s) get up-to-date, tailored prompts for the workshop. The content of the SME interview is not only used as a prompt, but is also used in the post analysis phase of the Threatcasting Method. See Sections 2.1.3 and 9.2.3 for more information about curating and interviewing SMEs.

Action

Based on your topic, write down a list of attributes an ideal SME might have.

1. What is the ideal expert's knowledge of the topic?

2. What field does the expert work in?

3. Has the expert written a book or articles about the topic? If so, what are they?

4. Is the expert a lecturer on the topic?

5. BONUS POINT: Does the expert offer a unique or opposing view from another expert you are highlighting?

Stories from the Lab

<u>Prompts in a Bucket</u>

The focused scope and nature of the small group Threatcasting can affect how the analyst(s) gathers prompts for workshop.

During Phase 0 for a small group Threatcasting Workshop that explored possible industrial and national security threats to Low Earth Orbit Satellites, the lead analyst(s) discovered a set of single recorded SME interviews was not sufficient to direct the participants. The reason stemmed from the specific focus of the Threatcasting Foundation. The analyst(s) were exploring the policy implications for these threats in a small group setting.

To better focus the participants, the lead analyst "bucketed" (grouped) SME recorded videos with policy specific articles and research papers. Each prompt then became a video plus a written piece. The combination created a sufficient set of four prompts, each a video and a paper, to focus the small group on the research question and application areas.

Other Prompts

When gathering prompts for a small group Threatcasting Workshop, the analyst(s) should remain flexible. Much like the analyst in the Story from the Lab (see above) the goal of the prompts is to direct the small group participants to explore the research question and apply it to the application areas (see examples in Section 2.1.3). Analyst(s) can combine multiple pieces of research in a single "bucket" or draft summary briefs that distill longer research papers.

Not all prompts need to be SME recorded video interviews. Using the different prompt categories (social science, technical, cultural history, economics, trends, and data w/an opinion {aka SMEs}) the analyst draws from a variety of material and sources to serve as prompts for the participants. Based upon the topic and research question the analyst can explore:

- online videos from conferences (e.g., TED talks, keynotes, book talks etc.);

- use excerpts from online video lectures, making sure they are not too lengthy and can be consumed by the participants quickly;

- book overviews and/or reviews;

- short articles that highlight more in-depth articles or papers. These can be found in periodicals such as *New Scientist* or *Journal Nature*. These publications will give a well-written overview that covers the main points of the longer academic or scientific paper;

- popular articles from publications (*Forbes*, *WIRED*, the *Economist*, *Scientific American*); and

- podcast or audio interviews, making sure they are not too lengthy and the participants can consume them quickly.

In addition, the analyst(s) can gather together multiple sources and create their own summaries and overviews. These can be presented to the participants as short written overviews or powerpoint presentations that capture the pertinent highlights of the research.

For a small group workshop, the ultimate goal is to gather just enough information in the prompt to get the participants to quickly work in their small groups with enough information to fuel their discussion and begin to answer the research question.

10.2.4   SELECT THE PARTICIPANTS

The lead analyst and core team select the participants to take part in the Threatcasting Workshop. If there is a steering committee, then they will help as well. Some of the participants will be new to the Method and some will be previous participants in Threatcasting events. The participants are involved in the Threatcasting process during the workshop and during the peer review of the final findings after the analyst(s) conduct the post-analysis. The participants should be sent the final report to share and have a copy of something they co-created with the team.

The curation of participants for a small group Threatcasting Workshop is important because of the smaller limited number. The task of selecting and inviting the people will take less time, but understanding the contribution of each participant will need more care.

How do you find participants?

The core team and steering committees come in when finding participants because they provide potential names. The team then searches each name and gathers unique data to determine their expertise and hobbies, creative practices, and other interests (these are generally from the bio links). This is also important to ensure there is a diversity of imagination in the room. Imagination is an essential tool, as is creativity, when building scenarios ten years into the future.

Creating threat futures with strangers demands a unique intersection of people coming together, requiring more than just different jobs. Finding participants for a small group workshop can rely more on your own social network and the connections of the core team and steering committee. The flexibility of the small group will mean that getting on people's calendars and holding the time logistically should be less complicated. But like the large group, try to get on the participants' calendars as early as possible, holding the time with a meeting notice or invitations.

Even with a small group, you will need to remain flexible when participants have personal and professional conflicts that come up. Because the group is small, a group coordination email can allow participants to "self-coordinate" by having a brief discussion around times that work for everyone.

Outliers and Outsiders

When building the participants list, it is also important to include those who are not inside the organization, system, or industry. Look for introductions to ethical hackers, science fiction writers, filmmakers, and poets as examples. Outsiders bring a fresh perspective to the topic and provide a unique lens into humanity.

Repeat "Threatcasters"

For a small group Threatcasting Workshop to run smoothly and efficiently, it is helpful to invite participants who have experienced a workshop before. Inviting repeat "threatcasters" can enhance the clarity of the EBM and general raw data collection because they already understand the Method and process. Ideally you will want one repeat Threatcaster in each small group as they can "coach" and encourage the other participants during the process.

Project Documentation

Even with the small number of participants, it is important for the team to track and document the participants. After the number of attendees is finalized, the next step is to place names on a tracking spreadsheet as a part of your Project Documentation. The names are organized by domain, expertise, gender, generation, and background. Columns are also included for links to online bios and other media. These links are a very important tool when building small groups too. The spreadsheet should be reviewed on the bi-weekly calls to determine where gaps appear, allowing the core team and steering committee to work together to fill them. Examples of gaps could be a shortage in a domain, gender, or diversity. You are trying to model futures and model threats so you can figure out how to empower people to make themselves and their surroundings more secure. To build these futures, it is ideal to invite as many diverse minds as possible to envision future scenarios.

Communications Plan

Generally, the communications that will need to occur with the participants are the simplest. At a minimum, they would include the following:

- a "save-the-date" invitation (once the date and venue are locked in);

- logistics information (at least two months before the event);

- Frequently Asked Questions sheet (at least one month before the event); and

- links to the workbooks (the night before the event so that they can pre-load them on their laptop/tablet).

These elements are discussed in Section 10.2.6.

Small Working Groups

The small group Threatcasting Workshop typically takes place over one to two days when it's in person, or it will span one to two weeks when it's virtual. This time is spent in small working groups of three to four participants. The analyst(s) and core team build the small groups. These small groups are where the majority of data generation happens during a workshop. Therefore, it is vital to have dynamic small groups to create the most robust raw data and narrative outcomes.

The concepts surrounding the curation of the small groups is discussed in Sections 2.1.4 and 9.2.4. The only difference between the large group Threatcasting and the small group Threatcasting rests in whether the small group workshop will be in person or virtual. If the intent is to run the Threatcasting Workshop virtually (and you intended to have portions of it asynchronous), then you should additionally consider the time zones of participants when putting together your small working groups. This will enable an easier time to schedule small group work if participants are in close time zones. More will be discussed about running a small group Threatcasting virtually throughout this chapter as tasks differ from the in-person concept.

Once the small groups are finalized with three or four participants, they are assigned a group name and will forevermore be identified by this name during the workshop and in the report. In the past the Lab has used colors for group names, such as Red Team, Blue Team, and Yellow Team. However, we have used items too: Team Boat, Team Bulldozer, and Team Plane.

Action

Prepare your supplies to build your small groups. You can accomplish this by using the tracking spreadsheet you built, a whiteboard, or a large paper sheet. If needed, gather dry erase markers, sharpies, phone, and computer charging cables.

Just a few days before the event, build your small groups. Don't do it any sooner as people will drop out of the workshop, and people could be added late. This activity is usually finalized the night before or even the morning of the workshop. Build your groups on your spreadsheet, whiteboard, or large paper sheet using these steps.

1. Create a grid of boxes equal in number to the numbers of small groups you will have at the workshop. As an example, 12 people would be 4 small groups of 3 participants each. In this example (see Figure 10.1 as a visual example), create four boxes with enough space to write in the names.

2. In these boxes, begin writing participants' names by their identifiers of a domain, expertise, gender, generation, and background to ensure every team has a unique and broad makeup.

3. Assign a color to each identifier. For example, military/government is blue, industry is green, academia (faculty and students) are red, and so on. Create a key to see the names appear as their unique identifier visually.

4. Prepare to move names around for hours, if not days. It is so important to create very dynamic small groups, so give this action a great deal of attention.

5. Once small groups are finalized, create a slide with all of the boxes, each including the small group participants' names and an assigned team name such as Team Red Pawn. This slide is used during the first morning presentation.

Team Top Hat	Team Shoe
Melissa Valero Thomas Harris Paul Royce	Sarah Grant George Cortes Bill McCormack
Team Thimble	Team Iron
Amanda Castro Nicholas Arnold Myron Dawson	Susan David Todd Smith Jeffrey John

Figure 10.1: Visual of participant names and team names for a small group Threatcasting (Primary background of participants: Blue is government, Red is academia, Green is industry).

Note that it is absolutely ok not to have each color represented in a box (Figure 10.1). That is because there are many characteristics you are interweaving into your small groups. In fact, many participants might span multiple primary colors. You just need to figure out what you are attempting to accomplish and how best to align the individuals.

The lead analyst's curation of the groups is meant to get the right participants together so that they will have an engaging and interesting conversation. People and personalities are not math. It is alright to have some overlap in backgrounds and skill sets. Sometimes this can't be avoided because of the number or mix of people. But having an understanding of these similarities and differences before you start the workshop will help the facilitator manage each small group appropriately.

For example, if one group has mostly academics, then the facilitator can encourage them to also take on an industry or government perspective. If a group has a gender or age imbalance, the facilitator can suggest to them that they also take on a perspective that is wholly different from their own. Knowing these subtle qualities before the event will help the facilitator communicate and ultimately get better raw data for the analyst(s).

10.2.5   DRAFT THE THREATCASTING WORKBOOKS

The Threatcasting Workbooks are the main tool used to capture the data that will be needed for the post analysis. The workbooks give the participants a place to write down their thoughts and perspectives on the prompts and are a place to capture the EBMs, which are the main output of the Threatcasting Workshop.

Analyst(s) should consider the Threatcasting Foundation as well as the makeup of the participants and small groups when drafting the Threatcasting Workbooks.

The collection tool can be a simple structured data gathering tool or a heavily designed (e.g., language, visual design) tool. Technically, the workbooks are a structured database, which is why simple spreadsheets work well for information capture. The workshop spreadsheets (workbooks) are used by the analyst(s) in the post analysis or as an input to other data processing platforms.

For ease of participants' access and version control, the Threatcasting Lab generally uses Google Sheets. The spreadsheet format is easy for structuring the data collection. Each small group has their own tab (which also allows each group to see what every group is writing in real time), as seen in Figure 10.2. Additionally, by hosting the workbook on Google Sheets, each participant can easily log into the document during the Threatcasting event.

Figure 10.2: Visual of small group tabs on the bottom of spreadsheets

For a Threatcasting Workshop, there are two types of workbooks to build in advance. These workbooks are not given to the participants until the night before or the morning of the workshop. This is because we want them to come with their experiences but without thinking about the problem set until they are in the room. The spontaneity and slight unknown for what is expected of them adds to the energy and creative output of the workshop.

Research Synthesis Workbook

The RSW is used with the prompts to capture the participants' analysis of the curated inputs. The RSW uses a series of open-ended questions to draw out the participants' perspectives and opinions. There is no fundamental difference between an RSW for a large group Threatcasting and a small group Threatcasting. For a refresher on the Method behind building an RSW, visit Section 3.1.2.

Action

Build an RSW (see example in Figure 10.3) in your preferred spreadsheet platform.

- Use open-ended questions to build your RSW.

- The goal is to have the participants explore and discuss the prompts, processing the information.

- Typical RSW questions include:

  ○ What was a data point that you found important or interesting?

  ○ What are the implications of the data point on the threat futures or research questions?

  ○ Are these implications positive or negative?

  ○ What should we do about it?

- Highlight the cells in a color (we use a pale yellow) to indicate where participants are supposed to be entering data. It does not matter the color (just something that shades differently so that it is also obvious to any color blind participants), this will just be a visual cue that they have work to do.

Data Point #	Prompt A Summary of the Data Point	Implication	Why is the implication Positive or Negative?	What should we do?
1				
2				
3				
4				
5				
6				
7				
8				
9				
10				

Figure 10.3: Visual of blank RSW.

Threatcasting Workbook

The second workbook to build for the workshop participants is the Threatcasting Workbook (TCW). The TCW contains a series of questions and tasks that lead the participants through the design and modeling process to develop their EBM. This process uses the tools of Experience Design and allows the participants to explore a person, in a place, experiencing a threat—all based upon the prompts and the RSW information.

Action

Build a TCW in your preferred spreadsheet platform. For a refresher on the methodology behind building a TCW, visit Chapters 4 and 5 (which include visual representations of these concepts). The TCW should contain the following.

1. Places for the participants to include their team name and a place to name their future.

2. Space to include the selected data points from the RSW.

3. Space for the participants to describe their person in a place experiencing the threat.

   a. Consider what extra questions might help the participants to tell a richer and expanded story. What would fill out the vision and give the analyst(s) more information for the post analysis? Use the Threatcasting Foundation. The research question and application areas can give specific guidance and language for more explanation.

4. Add experience questions using the experience design approach.

   a. Again, refer to the Threatcasting Foundation's research question and application areas as an initial prompt for where participants might give additional detail.

   b. Use The Six Dimensions of Experience Design (Experience Questions) to explore different aspects of the experience questions (Breadth, Duration, Interaction, Intensity, Design, and Value).

5. Add enabling questions to explore what factors helped to bring the threat about using the EBO approach.

   a. The Threatcasting Foundation's application areas will give the analyst(s) areas to consider for more questions for the workbook. These questions should highlight various enabling areas or activities.

      i. What economics or business need to be in place?

      ii. Is there new research that needs to be conducted?

      iii. Is there a specific failure of policy, law, or culture that enabled the threat?

      iv. Are there geopolitical conditions that contributed?

    b. Each of these enabling areas can help the analyst(s) expand and specify areas where they might want the participants to comment.

6. Add in the backcasting space for gates, flags, and milestones

    a. Are there specific questions you can ask that will help give the Who in your Application Areas more specific actions to take when they apply the findings?

Whether a small group Threatcasting or a large group Threatcasting, the TCWs will generally look the same. Figure 10.4 provides a snapshot of the top portion of an TCW. Figures 4.2–4.5 in Chapter 4 also provide visuals of a typical TCW.

Estimated Date of the Threat:	2029
Data Points	
NOTE: Roll the Dice to pick a data point from each of the research areas in the Research Synthesis Workbook (the rollup for each "SME Grouping" or topic)	
Grouping 1	attacks on critical infastructure
Grouping 2	Generation born into curated, personalized content have no resistance, resilience or immunity
Grouping 3	No existing playbook on privacy and the use data (esp. international agreements!)
Threat Actor or Adversary	
NOTE: Roll the Dice to pick a Threat Actor or Adversary (generally categorized by motive): 1) State Sponsored 2) Proxy	

Figure 10.4: Visual of a beginning (top) of a TCW.

10.2.6 LOGISTICS

The logistics for conducting a small group Threatcasting Workshop can be quick and manageable because of the number of participants. The following is a set of tasks to help the analyst(s) and core team conduct the workshop.

Decide on Modality (Physical or Digital?)

The first decision (that will drastically change the required logistics) is to decide whether you intend to hold the small group Threatcasting as an in-person event, synchronous virtual, or hybrid virtual event.

An in-person event looks very similar to a large group Threatcasting Workshop but could span either one or two days (depending on the experience of the group with the Threatcasting process). A synchronous virtual workshop means that all participants will be virtual but participate at the same time. This could look like 3- to 4-hour blocks of time over 3–4 days, or something similar. The third modality is a hybrid virtual workshop. All participants are virtual, but the schedule will consist of both synchronous and asynchronous times that would be spread over one to two weeks. For instance, you could schedule full group sessions for an hour every other day and allow each small working group to decide the best time for them to meet in-between to conduct the work.

Planning Meetings

The kick-off of a Threatcasting Workshop happens when the core team decides to run a workshop to gather the raw data to answer the Threatcasting Foundation. Planning meetings are generally all done over conference calls and/or video-calling platforms. For each small group Threatcasting Workshop event, the analyst(s) and core team will want to schedule bi-weekly calls for the first few months, then weekly calls the month before the event.

Action

For the calls, do the following.

1. Select a team member to be the call coordinator. This person will email the link to the conference call number to bring everyone together.

2. Set the agenda each week to move the project forward and bring everyone together.

3. Send out the agenda for the call 24–48 hours prior.

4. Select a team member to be the meeting host or facilitator to keep the topics moving.

5. Select a team member to be the scribe or note taker for the call to capture the discussion as well as the due outs.[19]

6. Follow up with any due outs after the meeting within 24–48 hours.

Pick a Date

The next agenda item is to determine the dates for the Threatcasting Workshop on the first kick-off call with the core team. Once that is established along with the modality, everything else can move forward. That seems like a simple action, to pick the date, but it never turns out as planned. Typically, the date is chosen at least 3–5 months out because many participants will have packed schedules and need that much lead-up time to clear their schedules.

If the workshop will be in-person, see Section 9.2.6 for more ideas of potential life events to schedule around. If the workshop will be a virtual event, picking the dates may seem easier, but holding on to those dates can be more difficult. Like for the large group Threatcasting, personal

---

[19] "Due outs" are all the things that need to get done before the next call. These tasks should always have a name attached to them so that individuals can report in on progress at the next meeting.

and professional conflicts arrive. Because the event is virtual, the "HOLD" that people have on their calendars may not stick as firmly as an in-person event. Flexibility is key. Additionally, if your small group is small enough it might be possible to use an online pole (doodle), email thread, or scheduling software (e.g., Calendy) to help find or adjust times.

Location Selection

As soon as the core team selects the date, the workshop's location should be selected. For the in-person modality, the core team must also come to a quick decision on the general split of participants (and their home locations) in order to decide on the best location for the Threatcasting Workshop.

Action

Things to consider when selecting an in-person location.

1. Is travel possible for all attendees?

2. What is the travel time and distance for most participants?

3. What time of year is it? This matters because those in cold climates would almost always prefer to take 2–3 days in the warmer west as opposed to vice versa. The time of year can also determine hotel costs.

4. Beware of locations with a negative connotation for some of your participants. For example, government employees have a hard time justifying attendance at events in Las Vegas based on the nature of the city.

5. Pay specific attention to any location restrictions that the Threatcasting sponsor has levied on them by their organization.

Venue Selection

Whether an in-person or a virtual workshop, a venue must be selected next. For a virtual workshop, the venue is the collaboration software that will be used for the workshop meetings. This could be Microsoft Teams or Zoom or something else. Preferably the software platform will facilitate both the full session meetings as well as breakout rooms for the small working groups. It is important for the core team to consider any restrictions that participants might have using any of the various platforms from their work or home locations (depending on where they intend to participate from).

For an in-person workshop, the venue selection process is identical to the process discussed for large group Threatcasting. However, obviously, a much smaller space is required and, therefore, the assumption is it will be easier to find and less expensive. But the core components are still required.

Things to consider when analyzing potential workshop venues.

1. Does the venue have enough space to host the number of participants you plan to attend the workshop?

2. Is there a side room participants can duck into for calls? Life never completely stops for a participant and sometimes they will have to take that work call. Therefore, having a semi-private space they can duck into is something nice to have, whether for a large group or small group Threatcasting event.

3. Is there an overflow space where participants can spread out during their small group sessions if they need a change of scenery?

4. Are there shops and facilities nearby to grab a quick coffee or lunch during a break?

5. Is there Wi-Fi that is easily accessible for all participants? Will that be an additional charge?

6. Are there enough power outlets at each work area for the small groups? Assume that each participant will want to plug in their phone and laptop/tablet.

7. Is there audio-visual equipment available in the main Threatcasting area? You will want the ability to project the facilitator's laptop with audio capability for the presentation of the prompts.

8. What are the costs of the facility rental and how far out can dates be locked in?

9. Are there any restrictions on the time that the facility is available? Namely, you want to make sure that you can access the facility in the morning at least an hour before the start of the event and host the event, un-interrupted, during the day. Generally, we try to reserve facilities from 7:00 a.m. to 5:00 p.m. on the event days.

Generally, the event space needs to have the following:

- enough desks/chairs in the main room to facilitate arranging the tables in small groups.

- couches or other table and chair seating, so small groups can move outside of the main room if desired.

- screen, projector, and sound system in the main room;

- IT support available;

- a nearby small conference room in case participants need to take a private call during the workshop;

- plenty of parking or easy transportation options nearby;

- food options nearby for pick up or delivery;

- a reasonable and affordable price within the event budget; and

- free Wi-Fi and/or included in the space rental fee.

Action

For an in-person event, do the following.

1. Contact rental venues to see what kind of packages they offer to get a sense of what is available in the area where the workshop will be held.

   a. Ask the core team and steering committee if they have ideas on viable event locations (and they might have the ability for a discounted rate).

2. Create a tab on your tracking spreadsheet called venue options. List the name of the venue, contact number, website, contact person, and link to the event package. This tab can assist with venue selection during core team bi-weekly meetings.

3. If possible, coordinate a time to walk through the potential venues taking pictures in order to share with core team members.

Build a Save-the-Date Announcement

Just like the large group Threatcasting, it is important to quickly (once the date and modality are decided) inform potential participants about the event so they can decide their interest and clear their schedules. The same level of attention to detail and personal touches are needed for a small group Threatcasting. Creating a digital postcard to send (with the workshop title, date, modality, research question, and appropriate/related images) will peak potential participants' interest. Figure 10.5 shows an example of a save-the-date graphic used previously in a small group Threatcasting.

Figure 10.5: Save-the-date announcement (virtual, small group Threatcasting Workshop).

Registration

The lead analyst or core team will also need to make a decision on how to handle registration. On one hand, you could use a similar registration process as used for the large group Threatcasting (see Section 9.2.6). This could increase the feel of professionalism for the event, plus it is just fun to get an event ticket (from using a commercial registration platform). However, given the small number of participants, it is also easy to assign a core team member or the lead analyst to manually invite and "register" participants.

Action

- Review your participant or potential participant list to determine the level of formality that might be needed for your invitation.

- Do you or the core team have a social connection to each person? This might make registration unnecessary and too formal.

- What is the total number of participants you are planning to attend the workshop? Is it small enough that an email or calendar invite might be sufficient to invite them and hold the time?

- Are most of the participants unknown to you? Are they located in multiple places or countries? Are there positions such (e.g., government, high-level corporate executive) that a more formalized process would make them feel more comfortable with the workshop?

Logistics Information

General logistics information should be sent to the confirmed attendees as soon as possible. Typically, with a small group workshop, most of the participants are local. But if participants are traveling from out of town, the logistics information will aid them in booking their travel plans and lodging accommodations.

At minimum, you should consider including the following information:

- address of the venue for the Threatcasting;

- public transportation options to arrive at the venue;

- closest parking garage options (with cost if available) to the venue;

- local lodging accommodations information (especially if a hotel block has been saved at a discounted rate for Threatcasting participants);

- closest airports to the venue for out-of-state participants; and

- rough schedule for the two days (so that participants can plan their travel accordingly).

The logistics information is generally reduced to the schedule and the venue platform if you are conducting the small group Threatcasting in a virtual modality. You might also offer to do a tech test session with any participants that are unfamiliar with the virtual platform you are using for the event. This will ease any potential disruptions to the first day of the schedule if people are having issues connecting.

FAQs

During the registration process, you should also provide participants with a FAQs list. This will ensure the participants know what is required of them. The FAQs list can be sent as a confirmation of registration. Having these details allows the participants to show up relaxed and ready to step ten years into the future.

Action

Build an FAQ, which could include some of the following items to help address the culture and execution of the event:.

- Will the small group workshop be physical, virtual, or hybrid?

  - If Virtual:

    - Explain the selected virtual meeting space, including the platform and how files will be shared.

- Offer a tech testing session if any of the participants are not well acquainted with the platform.

- Explain the schedule.

- How will you prioritize participants' time? In a virtual event generally the interaction between the people is the most important part. Consuming content can be accomplished alone by each participant.

  ○ If Physical:

    - Participants should dress comfortably and casually. The Threatcasting Lab's signature line is "Flip-flops encouraged" because we want everyone comfortable for the two days.

    - Please do not wear a suit or a uniform.

- You will do a mixture of large group and small group work. Small teams of three to four people are pre-assigned by the Threatcasting team. The assignments are provided once all participants are in the room.

  ○ No, you cannot find out who is on your team before the day of.

  ○ No, you cannot request to be on a specific team.

- Teams stay together and work together during the entire two-day event—everyone works; no one observes.

  ○ No, you cannot visit and watch the process. This sausage-making takes all hands on deck.

- There will be breaks for phone calls to do a quick check-in.

  ○ Your small team needs you, so do not plan long conference calls during the workshop.

- This is a working event; please bring a personal laptop that can access Wi-Fi and access the sharing platform that you are hosting the workbooks on.

  ○ You will receive the links the day before the workshop.

Accommodations and Transportation

Unlike with a large group Threatcasting, there are (typically) few individuals that will come from out of town (if doing an in-person event). Therefore, there is no need to reserve a block of hotel rooms near the workshop site. Instead, it is a nice courtesy to provide a list of hotels that are in walking distance to the venue to any out-of-town participants just to make their experience preparing for the event easier. Potentially also provide distance from airport or train station to the workshop venue and a list of all possible ground transportation including, rideshares, cabs, light-rail trains, bike, and scooter rentals—especially if any of the out-of-town participants are not familiar with the location. Include this information in the FAQs and/or logistics information email as discussed previously in this chapter.

Project Documentation

The established communications plan will be useful when scheduling core team meetings, sharing notes, and creating outcomes for the core team. Generally, the same planning documents are needed for both large and small group Threatcasting Workshops. However, they will be less complicated for a small group Threatcasting.

Action

Create a spreadsheet on a sharing platform such as Excel, Teams, or Google Sheets with tabs as the following.

1. Potential SMEs Tab with columns to denote name, affiliation, gender, domain, background, bio links, and status such as accepted, recorded, and transcribed.

2. Potential Attendee Tab with columns to indicate name, affiliation, gender, domain, background, bio links, and status such as invited, registered, and attending.

3. Seat Matrix Tab to create small groups and ensure the diversity of attendees. Columns denote name, affiliation, gender, domain, background, bio links, previous attendee, and recommender's name (see Figure 10.6).

A	B	C	D	E	F	G	H	I	J	K	L
First Name	Last Name	Organization	GOV	ACAD	IND	MIL	Gender	Speciality	Email	Link to Bio or Profile	Recommended by
Sarah	Jones	US Army				x	F	strategic planner	xxxx@xxxx		
Tim	Martin	MIT		x			M	PhD in AI	xxxx@xxxx		
John	Moses	Target			x		M	CFO	xxxx@xxxx		
Allie	Smith	Treasury Dept	x			x	M	financial crimes analyst; previous Army Ranger	xxxx@xxxx		

Figure 10.6: Threatcasting Seat Matrix.

4. Team "TO DO" Tab is a spreadsheet to track bi-weekly do-out tasks (see Figure 10.7). Many of these will be logistical tasks dealing with the venue and the "care and feeding" of the participants.

Rank	Item	Status	Next Steps	Owner	Links
1	Tech SME	Found 4 possible short videos	Decide which one to use for the event	John Robins	all saved in our drive
2	Core Team Meeting	June Meeting Complete	Send calendar invite for July meeting	Samantha Smith	
3	Save the Date Announcement	completed design	Send to potential participants	Samantha Smith	

Figure 10.7: Threatcasting TO DO tab.

5. Venue Options Tab is a listing of all the venues and their assorted information that you have researched within the city where you plan to host the Threatcasting workshop.

Planning a No-Host After-Hours Gathering

Whether a large group or small group Threatcasting, it is important to schedule an activity to allow participants to better connect with each other and their small working group as this will also build trust, which will yield richer threat futures moving forward.

For a virtual event, this is fairly difficult because they are not a captive audience in a location together. Instead, you could offer a happy hour time (after the first brief out of their EBM) on the virtual platform. This would be a time for social interaction and to get to know each other a little better. However, attendance will potentially be a little low since they are surrounded by other commitments like work and family that might not facilitate their attendance at the optional event.

For an in-person small group Threatcasting Workshop, we would schedule a no-host after-hours gathering at a restaurant within walking distance from the Threatcasting venue. With many of the participants (if not all) local, this may or may not be of interest to the group as many will have family commitments to get home to. However, if it does work, this will create more of the "summer camp" feeling you want to create, and when participants come back the next day, their trust in each other has grown, and they do a deeper data mining of themselves and each other.

Randomness Equipment

Recall that in the participants' construction of EBMs, they will need a random source to select the data points from the combined RSW. If you are conducting the workshop in person, you will need to procure enough equipment (dice, playing cards, etc.) for each small working group to have one set. If the workshop will be virtual, then you will need to figure out an internet-based pseudo-random number generator to use or to pick a physical object that participants should generally have in their house.

Prepare Equipment

Whether conducting the workshop virtually or in-person, it is important to capture a detailed list of all the equipment that will be used/needed during the event. Once the list is created, then the core team and/or lead analyst will need to pull it all together and test its functionality before the event.

For an in-person event, this could be items like:

- IT equipment in the room (i.e., screens, Wi-Fi, speakers, microphone);

- electrical outlets, extensions cords, charging towers;

- timing app (for the report outs);

- facilitator's PowerPoint slides; and

- SME videos and other prompts.

For a virtual event, the equipment is also virtual:

- virtual meeting space with break out rooms;

- collaboration platform to share video and documents;

- facilitator, microphones, lighting (if needed); and

- stop watches (physical or app).

Catering

For an in-person small group Threatcasting Workshop, catering is not always a requirement. Sometimes groups opt to take a break and grab a coffee or lunch on their own. However, if you chose to bring in food to save time, the requirements are identical to a large group Threatcasting Workshop, although the numbers are significantly lower. See Section 9.2.6 for a detailed explanation on how to handle food requirements for the event. For a virtual workshop, there is no requirement for food.

Check-In

If possible, access the event space the evening before the workshop to ensure set-up details (such as tables and chairs, projectors, screens, and space for catering) are complete. This way there won't be as large of a scramble in the morning.

This is our morning of checklist for an in-person event (which is significantly more informal given the size of the group and our connections with them).

- Designate a check-in table where a teammate can sit in order to greet participants as they arrive. This should be right outside the event space, clearly visible from the entrance to the floor.

- Place the final registration list on a clipboard (bring pens), laptop, or tablet (don't forget the charger and stylist) to check participants off the list as they arrive.

- Sort participants' name tags in alphabetical order on the table surface for quick pick up as the participants check-in.

- Hand participants their name tag and small group assignments. These should be in the form of colored game pawns or mini plastic toys. Ask them to hold on to the objects for the morning.

- Let participants know that for the opening morning session, they can sit anywhere in the room they want to. They do not need to find their groups yet.

- Inside the workshop room, tables and chairs should be pre-set for teams of up to four.

- Catering should also have a table inside the workshop space or conveniently nearby. For the morning, it should be ready with coffee, tea, and water.

- The screens should be ready for viewing, all presentations loaded and Wi-Fi turned on.

- Extension cords and charging stations should be placed throughout the room.

- Place bowls of candy on every table.

10.3   PHASE 1

In Part 1 of this textbook, we explained Phase 1 as the Research Synthesis portion of the Threatcasting Method. However, when discussing the application of the Threatcasting Method, you should visualize Phase 1 beginning at the start of Day 1 of the Threatcasting workshop.

10.3.1   WORKSHOP OPENING

Because of the small number of participants, the opening of the small group Threatcasting (virtual or physical) workshop can go quickly. However, it is important to take the time to create the creative environment that the process needs (regardless of whether the event is in-person or virtual). If it is in-person, then a similar opening as described in the large group Threatcasting would make sense.

Facilitator

Once the opening session concludes, the floor is yielded to the workshop facilitator activities. The facilitator is in charge of keeping the room's time, keeping the participants on task, and making the experience feel like a fun and creative workshop. A facilitator should have a dynamic personality and a boldness that allows for maintaining control of the floor (regardless of whether the floor is in physical space or virtual space).

Throughout the entire workshop, the facilitator should ask participants frequently if they need anything or have any questions; they will, and walking around, continuously connecting with them makes the process easier for all involved. The facilitator should frequently remind the participants that they are experts, and that's why they were invited to the workshop.

The facilitator's role becomes more challenging if conducting the workshop in a virtual environment. The facilitator serves the same function to explain the method, answer questions as well as encourage and prompt the participants. In a virtual workshop the facilitator or member of the core team will need to manage the virtual meeting space, taking on tasks like sending the small groups to their breakout rooms or meeting areas. Also, the facilitator will need to jump into these breakout rooms to make sure the participants have what they need.

When conducting a virtual workshop, the RSW and EBM workbooks should be a shared document whenever possible (e.g., Google Sheets) This will allow the facilitator to monitor the participants work and intervene where needed.

Workshop Agenda

Each day of the workshop should have an agenda built as a slide deck to provide instructions for the participants. The Threatcasting Lab's deck includes the following items.

- Introduction to the Threatcasting Process

- Relevant definitions within the problem space

- Reveal of the assigned Small-Group Teams

- Phase 1: Research Synthesis

- Phase 2 + 3: Futurecasting + Backcasting

If the workshop is in-person, then the timeline should generally follow the timeline for a large group Threatcasting, as seen in Section 9.3.1. However, the amount of time needed for report outs (from the RSW and TCWs) will be less given the smaller number of participants.

If the workshop is synchronously virtual, a sample timeline might look like this:

- Day 1 (1–2 hours): Opening session, Introduction to the Threatcasting Method, Phase 1—Research Synthesis

- Day 2 (2–3 hours): 1st EBM (Futurecast and Backcast)

- Day 3 (2–3 hours): 2nd EBM (Futurecast and Backcast)

• Day 4 (1–2 hours): Initial draft findings, Discussion

If the workshop is hybrid virtual, a sample timeline might look like this:

• Day 1 (1 hour): Opening session, Introduction to the Threatcasting Method

• In-between: Individuals and then small working groups execute Phase 1

• Day 3 (1 hour): Report out on RSW; Discuss Phases 2 and 3

• In-between: Small working groups execute 1st EBM

• Day 5 (1 hour): Report out on 1st EBM; Discuss changes for next EBM questions

• In-between: Small working groups execute 2nd EBM

• Day 7 (1 hour): Report out on 2nd EBM; Discuss results

• In-between: Analyst(s) conduct initial Phase 4 activities

• Day 9 (1 hour): Analyst(s) report out on initial draft findings, AAR

Workshop Presentation

Build the presentation for the workshop in advance (powerpoint or your preferred software). It's important to tell the participants what we're going to say during the facilitated opening and then say it, and then again as a way to over-communicate. And then on every day, tell them again what you already said. This helps set the expectations, and the repeated messages build familiarity and trust with the participants. We keep the PowerPoint presentation up on the screen during the entire day—whether it is a physical display in a physical room or in the main virtual room when they are in breakouts. This way we can either highlight the current directions for the task at hand or display the current timeline.

10.3.2   PRESENTATION OF PROMPTS

The prompts are presented to the participants in order to conduct the research synthesis activities. Similar to a large group Threatcasting, these prompts can be SME videos or "buckets" that combine videos with other research mediums. The major potential difference between a large group Threatcasting and a small group Threatcasting is that this phase can be assigned as "pre-work."

If pre-work, the participants will watch the prompts on their own time, fill out the RSW, and come to the workshop prepared to discuss. Alternatively, the small group Threatcasting can be conducted identically to the large group Threatcasting—this phase would be conducted during the workshop time. See Section 9.3.2 for more details and considerations when presenting the prompts.

However, whether the prompt viewing and individual research synthesis is conducted before or during the actual Threatcasting workshop, in both cases—it is still critical to have time in their small working groups to discuss and expand their individual RSWs in order to capture the "wisdom of the room."

Shortened Time Small Group Workshop

There are some circumstances when a small group Threatcasting workshop is small not only because of the number of participants but it is also small because of the amount of time that the participants can devote to the workshop. Often when conducting a large group workshop, gathering that many people together means that they will devote one and a half to two entire days to the work. You may have a similar time commitment from a small group Threatcasting but you may not.

One way the analyst can work within a limited time constraint is to change the agenda for the workshop and allow the participants to consume the prompts and fill out the RSW individually prior to the event. This requires that the participants commit to the pre-work. But if they do, then the workshop can begin with diving directly into the collaboration on the EBMs. Some analyst(s) will record a video of themself explaining the prompts and how to fill out an RSW so that the participants can review before doing the work individually.

Below is a sample agenda (for a shortened time small group Threatcasting workshop).

- Day 1 (1 hour): Opening session, Introduction to the Threatcasting Method, Explanation of RSWs, and prompts.

- In-between: Participants consume the prompts and individually fill out a RSW.

- Day 3 (1.5 hour): Quick RSW discussion; small working groups break out and fill out 1st EBM.

- Day 4 (1.5 hour): Report out on 1st EBM; discuss changes for next EBM questions; small working groups execute 2nd EBM.

- Day 5 (1 hour): Report out on 2nd EBM; discuss results and next steps.

10.3.3   DATA CAPTURE IN WORKBOOKS

At the start of each Threatcasting workshop, all participants are provided with a link to the workbooks, a shared document space, or digital files. The small working groups will use these to write their ideas, notes, and scenarios. Using workbooks allows more data to be captured than appears in the verbal report outs because no one can do justice to everything that was in their heads. The idea then for the workbooks is to get as much detail written down as possible. The workbook gives the small group space to start looking at the problem from various angles. As the workshop progresses, each small group begins to populate the workbooks' fields.

After reviewing the prompts, the small group participants break into smaller working teams and begin work on the RSW. Here they capture data points, implications and possible actions that draw from their own experience (see Sections 3.1.2 and 9.3.3 for a refresher).

A Note on Scribes

When the small groups gather to put information into the workbooks (e.g., RSW, TWC) the facilitator should encourage the small working groups to pick a scribe. This will be the person who will capture the conversation, discussion and disagreement that happen in the group. The scribe is essential to the Threatcasting Method. If the data is not captured fully in the workbooks it cannot be used later in the workshop nor will it be of any use to the analyst(s) during the post analysis.

Because the scribe is so important, the facilitator should check in with each group and identify the scribe. They should be thanked and told that they are an important part of the method. The facilitator should then encourage the rest of the group to also write in the workbook, to capture as much information as possible. High performing groups often will have multiple members of the small groups adding in information to the workbooks.

For a small group workshop that might take place over a single or multiple days, the facilitator should encourage the small groups' members to share the scribe task. A single person should not be the scribe for every step in the process if possible.

10.3.4   DISCUSSION

Throughout the course of the Threatcasting workshop, the small groups will "report out" multiple times to the other participants in the room. Small groups report out so that everyone in the room can hear what everybody else is thinking. During report outs, participants have a strict time limit to present the most important thing(s) that came from their previous activity. With respect to this first "report out," the small groups are typically asked to provide the top three important data points that came from their research synthesis activity, including the answers to the other research synthesis questions for those three points. The facilitator also needs a stopwatch to keep track of the time. When the timer goes off at the end of the allocated time (perhaps 2 minutes), we let the alarm go off in the room, so the reporter knows time is really up.

The main difference between the small group Threatcasting and the large group Threatcasting is the coverage of the prompts during the research synthesis activities. Namely, in a large group Threatcasting—you might have multiple small working groups that analyze a single prompt. In a small group Threatcasting, it is more typical that each prompt will only get one small working group to analyze it and sometimes a small working group will need to analyze more than one prompt.

Therefore, care is needed to determine which small working groups should analyze which prompts. Recall the discussion in Section 9.3.2.

Therefore, once all the small working groups have presented, it is important to open the floor to a short discussion to encourage participants to share what didn't get talked about—perhaps highlight some of the side conversations that they had in their small groups that were not presented. "What didn't come up" is an important element of Threatcasting because what isn't talked about often illuminates the things that should be discussed. You should have an analyst or a co-facilitator take notes on this conversation as the comments might be helpful in the post-analysis phase.

10.3.5 PREPARE FOR NEXT PHASE

Given that there was probably minimal overlap in the small working groups analyzing the same prompt, the "magic" transition of the data that participants provided in the RSW to the set of final data points that will be used during the rest of the workshop is fairly easy (compared to a large group Threatcasting).

Still, in order to maintain the consistency of the data, you should still create new tabs in the RSW workbook to consolidate the data points (per prompt) for the next phase. It also makes it easier for the small working groups to associate the data points with the prompt number vice the team name that did the analysis.

For purposes of illustration in Figure 10.8, Bucket A will hold the consolidated data points from prompt #1, Bucket B on prompt #2, etc. while still maintaining the initial work performed by each Team. You could also choose to add additional data points that occur during the discussion period if you desire into the appropriate Bucket A, Bucket B, etc. Recall that the purpose is to capture the "wisdom of the room."

Figure10.8: Visual of a combined RSW's tabs (post Research Synthesis).

See Section 9.3.5 for more information on this data management step.

Walking Breaks

Build in breaks during the Threatcasting process. During the multi-day workshop, participants are provided with places they can easily walk to that are filled with nature, interesting sites, or shopping. We do this because there is a lot of sitting and writing during a workshop, and people still need to move. We call these walking breaks, and use them, so that participants don't get stagnant during the exercises; we encourage participants to take a walk.

Even if a workshop is virtual, breaks are important. Build breaks into the agenda and encourage the participants to get up and walk about, get some water or maybe even do some light stretching. The virtual meeting space can be left open if participants want to hang out and chat during the break.

10.4 PHASES 2 AND 3

Futurecasting and Backcasting

See Chapter 4 in the methodology section of the textbook for a refresher on the EBMs that the participants will now construct.

Generally, the facilitator will walk the participants again through Phase 2 and 3 of the Threatcasting process using the TCW as the prop. Ensuring participants are comfortable with the workbook layout and location is critical to success. It is also important to have technical assistance standing by in case participants have issues accessing the new workbooks. In a physical workshop, this is pretty simple to coordinate access to an IT individual with the venue. In a virtual workshop, the suggestion is to have an IT individual logged into the collaboration platform and a "breakout room" set aside for technical assistance so that the participant that is having issues does not derail the entire group (instead they can move to the IT breakout room to get assistance).

10.4.1 CAPTURE THE RAW DATA

Picking the Foundational Data Points for the Workbooks

For many participants, picking the data points from the RSW for the EBM can be cumbersome and time consuming. Time is short when the participants are modeling their futures, therefore it's important to help them pick as quickly as possible.

Additionally, the goal is to have the participants use data points that could be challenging. The analyst(s) don't want the participants to cherry pick the data points they most prefer or know something about. It is ideal when the data points challenge the participants to consider and discuss how each of the data points interacts with the other. Often analyst(s) get the best results when the data points used by the participants conflict. Resolving that conflict yields valuable raw data.

Finally, a level of "randomness" is ideal for the participants to drive their EBM. The level of randomness depends on the method or tool provided to the participants. The method can be truly random or at the very least give the appearance and feeling of mild randomness.

There is a listing of different tools and methods in Section 9.4.1 that could be used (as it is similar for both the large group Threatcasting and the small group Threatcasting).

Building the EBM with "A person, in a place, experiencing a threat"

Getting started is always the hardest part for the participants, especially if it is their first time filling out an EBM. Often the groups want to pick the "right" person or get their threat future perfect before they begin.

The facilitator should encourage them to just "jump right in." This is generally easier in the small group Threatcasting as the facilitator can spend more time with each team. Giving their person a name generally helps. Having the group pick a specific city also helps to get them started and also narrow down the expanse of ideas. Participants can feel overwhelmed with the blank page so getting something written as soon as possible helps to break the ice.

As we explored in Chapter 4, the questions are meant to encourage the participants to tell the best story they can tell. In this way, the secondary/sensing questions can give more detail (e.g., What will it smell like when the person experiences the threat?).

Tailoring these questions right from the start, based on the Threatcasting Foundation, can expand the quality of the raw data in the EBM.

Experience Questions

Typically for a small group Threatcasting workshop, the analyst(s) will have crafted the experience questions in Phase 0 to be more specific than in a large group Threatcasting. Because the group is smaller, the diversity and experiences of the participants will not be as wide, and the total data gathered will be less. These more specific experience questions will allow for the small group to capture the various perspectives of all the participants in the group. Don't forget, you can always tweak these questions in successive iterations of Phases 2 and 3 during the workshop if they initially don't gather data that helps to answer the Threatcasting Foundation.

Enabling Questions

Like the experience questions, the enabling questions for a small group workshop can be more specific. The aim for this approach is different from the experience questions. The enabling questions in the EBM are used to specify what has happened or what is needed to bring the threat about. Rooted in Effects Based Operations, the domain expertise of the participants and the curation of the small groups takes on added importance in the small group Threatcasting workshop.

Again, the facilitator needs to remind the participants that their professional backgrounds matter but that they should also focus on the Threatcasting Foundation's Application Areas. The more specific they can get, pulling from real work experience, the more useful the raw data set will be for the analyst(s).

Often, asking clarifying questions can help.

- Is there a specific law or regulation that was or was not passed?

- What group, industry, or market might need to develop a specific technology?

- Are there specific areas of investment that needs to happen from specific organizations or countries?

- Have you seen something like this before but applied to a different threat or future?

Usually the participant(s) know more and have more detail in their minds than they are entering into the EBM. A few simple questions can help spark them to explain more.

Gates, Flags, Milestones

Backcasting is the final part of the EBM. This is the area where most small groups get bogged down, tripped up, or struggle to finish. First, this is usually a casualty of their time management. The groups will spend more time on the opening sections of the EBM and not save enough time to explore the gates, flags, and milestones. By the second day of the workshop (if there is one) the groups will have usually gotten better at time management.

The facilitator should monitor the progress of each group, encouraging them to save time for the final section. For a small group workshop, the facilitator can spend more time discussing the backcasting with the participants.

Also, it can be helpful to suggest that the group does not have to "go in order." Meaning, the small group can skip around the EBM, filling in different details and information that strikes them. Giving them the option to race through the EBM as quickly as they can to get to the end—followed by the option to revise, edit, and add—has been an effective strategy to get more detail and more robust gates, flags, and milestones.

Sometimes being vague is alright. Even if the participants don't have the specific expertise or understanding for how an action could and should be taken, they should write down as much information as they can. This can be a call out that more work needs to be done.

Example vague language that can be effective:

- a broader cultural conversation must be started to explore...

- academia needs to research the possibilities or perils of...

- an industry consortium needs to be convened to define the problem of ...

These vague approaches and suggestions can give specific organizations (e.g., industry, academic, etc.) a place to start. They also allow the analyst(s) to call out the needed actions in the post analysis or follow up with SMEs to get more detailed steps to be taken.

Report Out

Just like the report outs at the end of Phase 1 (Research Synthesis), each small group is given the main stage to share their EBM (vision of the future). Using a timer (perhaps 90–120 seconds) ensures that everyone has a chance to share in a concise manner. It is also a camaraderie building activity as they race to share their story about their person before the timer goes off. You should encourage the "storyteller" for the group to change each time there is a scheduled report out.

Given that they generally share more details than might be written down, it might be helpful to have the co-facilitator taking high level notes with an eye toward the post analysis. This is also a good point to remind the participants to provide more details in their written product to reflect all the good ideas that they have come up with during this activity.

There are really no differences between the EBM report outs between the large group and small group Threatcasting Workshops. The main difference between EBM report outs for an in-person virtual event is the creativity of the facilitator in displaying the count-down timer and how to cut the speakers off at time. The time still works as a great tool to keep the workshop on track. Holding up the time to the screen and calling out to the participant when their time is up generally will keep things going. Once you have done this once or twice, the participants will catch on to the flow of the workshop and stay within the time.

10.4.2 RINSE AND REPEAT

Typically, the futurecast and backcast process is repeated two to three times during the course of a Threatcasting Workshop, wherein the small working groups fill out new and unique TCWs for each round. The more rounds that can be accomplished, the greater the volume of raw data for the post analysis. By "unique" we mean that each small working group will select new data points from the prompts for each round.

Additionally, after each round of EBM creation, the lead analyst should modify some of the questions or add constraints to the participants. For instance, you could require that the person has to be in the military or the place as to be outside the U.S. or that the small working groups can use a technical data point of their own creation (pulling from the RSW). Needless to say, there are many ways to slightly change the foundation of the EBM in order to get different data. Mostly, these changes are to push the group in a different direction from their previous EBM to ensure that enough unique data is captured to cover the required Threatcasting Foundation.

Also, to keep all the EBMs straight (and help ensure that a group does not write over a previous product) you should create a new TCW for each iteration of Phases 2 and 3. This will also help with the post analysis to understand the new constraints imposed in each iteration. And, since the small working groups will remain intact, it will also allow the lead analyst to see their growth in ideas.

Timing

As with the in-person large group Threatcasting, the small group Threatcasting Workshop will follow a very similar schedule (as annotated in Section 9.4.2). If performing the small group Threatcasting virtually, there is some flexibility in timing. For instance, if it is an asynchronous event, then the participants will need adequate time (outside of the event) to create the EBM, so potentially leave 24–48 hours between report outs. This gives them time to handle life events and provides a better chance that the small working groups find adequate time in their schedules to meet. If it is a synchronous virtual event, the timings for the creation of the EBMs can be about the same as the in-person workshop.

10.4.3   CLOSING THE THREATCASTING WORKSHOP

At the close of the workshop, over-communicate what the participants can expect next. This will help to answer many questions now that they have given so much time and energy to the workshop. Ending the workshop with discussion, reflection, and gratitude will honor the participants' contributions and also encourage them to continue working with the analyst(s) during the post analysis and peer review phases.

After the workshop concludes, there are a few immediate actions to take. If the workshop was physical, then these will be focused on returning the venue space back to its original condition. See Section 9.4.3 for more details.

If the workshop was virtual, it is much simpler. Make sure you have backups of all the workbooks and notes from the workshop. If there are any documents left in a shared meeting space, remove them.

10.5   PHASE 4

After the completion of the Threatcasting Workshop and the generation of the raw data, the next phase in Threatcasting is post-analysis.

10.5.1   ANALYST(S)

Because the volume of raw data developed in the small group Threatcasting Workshop is significantly less than is developed in a large group Threatcasting Workshop, the lead analyst can generally complete the post analysis. However, to get different perspectives it can be helpful to bring in additional analyst(s) for review. Typically, each analyst will work alone on the three steps of post analysis: summary, meaning, and novelty. Once the three steps have been executed, the analysts convene and present the results. Regardless of how you structure your analysis cell, it is still important to create a timeline for your actions.

Action

Build a Timeline

The analyst(s) will need to craft a timeline for the post analysis that supports the required completion date of the final products.

1. Coordinate with all analysts on the project on their availability.

2. Determine whether to conduct the post analysis in-person or remote.

3. Schedule a first review session to take place the one to two days following the workshop.

4. Provide ample time for the Analysts to conduct their individual analysis of the raw data.

5. Convene analysts together for collaborate sessions periodically through the process.

6. Draft findings.

7. Establish reviews of the findings with the core team and participants.

8. Finalize findings.

9. Present findings to steering committee.

10. Begin work on final output.

Post-Analysis Workbook

The post-analysis workbook is a part of the project documentation, typically a spreadsheet. It is a place for the analyst(s) to capture each round of the post analysis (summary, meaning, novelty). It is beneficial to use this so that it can be shared with other analysts and also so that it can be included in the final output, assuring transparency in the methodology. Therefore, just like the other workbooks used throughout the Threatcasting Method, we use Google Sheets for the post-analysis workbook.

There is no difference in the post-analysis workbook from a large group Threatcasting or a small group Threatcasting. A visual of this workbook is seen in Figure 9.10 in Section 9.5.1 and other examples are found in the Project section of the textbook.

10.5.2   PREPARE THE RAW DATA

Data

As stated previously, the output from a small group Threatcasting Workshop is small and therefore, more manageable. However, you still need to devise a data management and data protection plan to maintain the transparency of your analysis work.

Data Hygiene

Phase 4 starts with data hygiene on all the data in order to prepare it for the post analysis. Chapter 6 in the methodology section of the textbook can provide a refresher on these steps.

Action

1. Immediately following completion of the workshop, ensure that you have a complete copy of all the RSWs and EBMs saved in an off-line (but protected) location. This will ensure that data can be recovered if it is accidentally corrupted during Phase 4.

2. Write down your data management and storage strategy and share with the analyst(s).

3. Once the assigned team member conducts the data hygiene step, create a complete copy of all the RSWs and EBMs and save it in an off-line (but protected) location.

10.5.3   PERFORM THE ANALYSIS

Post Analysis

First, there is no significant difference between how to perform the post-analysis phase for a small group Threatcasting and a large group Threatcasting.

Typically, each analyst will work alone on the three steps of post analysis: summary, meaning, and novelty. Once the three steps have been captured, the analysts convene and present the results to facilitate a group analysis on the raw data. Generally, this can be done over the span of 2–3 days (depending on availability of the analyst(s)).

Every analyst has specific tools they prefer to use and a preferred environment to work in. However, it is important to capture their analysis techniques and results.

Action

1. If there is only a single analyst: Review your notes and any project documentation that would help give an additional perspective before you begin.

2. If there are multiple analyst(s):

a. Document and discuss each person's expertise and how they are thinking about approaching the raw data.

b. Set a timeline for each analysts' solo post analysis.

c. Select a shared project documentation template so that each analyst is using the same format in the end.

3. Write down the Threatcasting Foundation on a sticky note or piece of paper so that you can refer to it throughout the process.

4. Periodically step away from the data to clear your head.

5. Remember to use the techniques (described in Chapter 6) for each round.

6. Keep notes in the project documentation for any insights, ideas, or early clusters for yourself and other analysts.

10.5.4 GENERATE, VALIDATE, AND REVIEW THE FINDINGS

Generate

Following the three steps of post analysis as described in Chapter 6, the analyst(s) will generate their findings. Using the Threatcasting Foundation as a guide, the analyst(s) reviews Round 3 "novelty" and considers how they answer the Research Question as it is applied to the Application Areas.

It is important to remember that it is the lead analyst's role to have an opinion based upon the initial research and SME interviews when combined with the raw data sets from the workshop. Due to the smaller size of the data set generated in a small group Threatcasting, often the lead analyst will have to perform secondary research/interviews to flesh out the findings.

Validate and Review

To validate the findings, the analyst(s) first explore if the findings answer the research question and application areas. Next, the analyst returns to the initial research and SME interviews to validate the findings and also identify if there are any outliers or missing perspectives.

Integral to the validation and review of the findings for a small group workshop is communication with the core team, steering committee (if it is being used), participants, and SMEs.

10.5.5 COMMUNICATION

Core Team and Steering Committee

It is important to keep the doors of communication open between the analyst(s) and the core team/ steering committee during the post analysis. Confirm with the various core team members how they wish to be involved in the post analysis. Some core team members are curious and will appreciate updates throughout Phase 4 and others prefer to see the draft results at the end of Phase 4. Also, let the steering committee (if you are using one) know when you expect to have some raw results of the post analysis for their feedback so that you can schedule an update meeting. Ultimately, you might ask these groups to conduct a peer review of the initial findings and to also sign off on the final report before it goes public.

Participants

It is also important to maintain communications with the participants during the post-analysis phase. First, we want to give them the opportunity to share any closing thoughts about the venue, process, and event. Next, we need to ask them how they want to be attributed on the final products. Some participants are fine with listing their name with organizational affiliation, and others prefer to remain anonymous. Also, many participants will have multiple affiliations, so this is the opportunity to determine which one they want to use. Finally, the participants should be given an opportunity to serve as peer reviewers on the findings.

10.6 PHASE 5

Similar to Phase 4, there is no significant difference in how Phase 5 activities are conducted between a large group Threatcasting and a small group Threatcasting. Therefore, see Sections 7.1 and 9.6 for more details.

Action

1. Review your Threatcasting Foundation to determine how your sponsor wants to receive the findings.

2. Review Sections 9.6 and 7.1.1 for more details on output types.

10.6.1 TRADITIONAL OUTPUTS

Technical Reports

Once again, this is the staple of outputs from a Threatcasting workshop (no matter the size). Section 7.1.2 has more details on content and scope for a technical report. A professionally written report is a typical deliverable to show the results from using the Threatcasting Method. As you are working through the phases, keep this output type in the back of your mind. This way you can think about the graphics/photos you might want to incorporate into the final technical report as well as which of the participants' EBMs you might want to turn into narrative form as part of the report's design.

Action

1. Create outline (sample outline is provided in Section 7.1.2).

2. If more than one analyst or writer, divide up initial writing responsibilities.

3. Write first draft.

4. Select graphics to be used that will enhance the report sections.

5. Select participants' EBMs to use that highlight a finding.

   a. Turn those EBMs into a short narrative.

6. If more than one writer, have each writer review/edit the other sections in order to create the second draft. If only one writer, review the draft (with the graphics/EBM components) and revise.

7. Share this second draft with participants and members of the core team for their feedback and comments.

8. Review the provided comments and incorporate into the technical report as desired.

9. Have an outsider (i.e., someone that has not read any aspect of the report) conduct an editorial review to catch tone inconsistencies and grammar.

Academic Papers

For a subset of the findings from the small group Threatcasting, you might want to write an academic paper, especially if one of the gates is the requirement for more academic research on a specific topic. Publishing a paper or short article in an academic journal within that discipline could be very helpful for disseminating your findings. Also, if you are a student or a new analyst, this could be the start of your publication history within your curriculum vitae (CV).

Writing for academic publications might be a slightly different style than you are used to. Therefore, it can be helpful to find a writing mentor to help guide you through the process.

Action

1. Write down novel ideas and findings that might work as the foundation for an academic paper.

2. For each idea, brainstorm different participants and SMEs who might be good collaborators for the academic paper.

3. Locate a writing mentor to help with the process, especially if academic writing is new to you.

4. For each potential idea that you listed in #2, brainstorm a list of possible publications that might be appropriate for the paper. Record the following information on these publication venues:

   a. Publication Name

   b. Website

   c. Submission Deadlines

   d. Submission Criteria (including length, format, etc.)

5. Select one of the ideas and create an outline (that confirms with style guidance from the publication).

Briefings, BLUFs, Executive Summaries

A BLUF (Bottom Line Up Front), a summary, or a briefing offers a window into the high level of the analyst(s)' findings. The analyst(s) should always plan to create a briefing that summarizes the effort. Generally, you should plan to offer this briefing to any person or organization that has been involved in the project. Aside from being a professional courtesy for participating, it is also an excellent way to socialize the findings and increase the social network for the analyst(s).

See Section 7.1.2 for more details on how to construct these output types.

10.6.2 ALTERNATIVE OUTPUTS

Graphic Novellas

A graphic novel is a story presented in comic-strip format and published as a book. A graphic novella is a short version of a graphic novel. Those who do not have the time to read the entire technical report, nor a lengthy academic paper, will likely have the time to read a ten-page graphic novella. Typically, this will be enough to whet their interest in starting conversations about the topic within their community or organization or reading the other papers. Section 7.1.3 has an example of a graphic novella generated from Threatcasting results.

These graphic novellas can be highly impactful because they can show the reader what a possible or potential threat future could look like. Because it is a visual medium, the writer and creators of the novellas can pack in a lot of information that was developed in the workshop and written about in the technical report. Section 9.6.2 has additional details and actions to help you determine if producing a graphic novella on a Threatcasting finding would be useful.

Podcasts, Interviews

Finally, an interview and/or a podcast might be an effective way to provide an audience with a high-level overview of the Threatcasting findings (similar to a written executive summary)—just in audio form. See Section 7.1.3 for examples of interviews and podcasts on Threatcasting findings. As recording a podcast or giving an interview is the same for a large group Threatcasting or a small group Threatcasting, see Section 9.6.2 for action steps to prepare for and execute this audio events.

CHAPTER 11

Individual Threatcasting

"If you know you are on the right track, if you have this inner knowledge,
then nobody can turn you off"
—Barbara McClintock, cytogeneticist Nobel Prize Physiology or Medicine 1983

11.1 INTRODUCTION

An individual threatcasting is just as it sounds; the analyst is also the participant. Moving through each of the phases as a dual analyst/participant allows the individual more freedom to curate prompts and adjust the EBMs to produce additional data. This curation helps define the intersections that the analyst will be Threatcasting within and gives a framework to create multiple EBMs, with a person, in a place, experiencing a threat.

The advantage of doing a Threatcasting as an individual is less coordination (because it is just one person) and there is more control over the timeline. Individual Threatcasting allows the analyst to be more nimble, to adjust, to explore new inputs and iterate on the Threatcasting process to generate more data and findings that will be more helpful.

A disadvantage of performing an individual Threatcasting is the lack of perspectives that a single individual provides to the problem space. In both the small and large groups, multiple people work together, have discussions, explore ideas, and collaborate. This collaboration allows people to come up with perspectives that they would not typically have by themselves. Diverse backgrounds and influences create a more abundant data set to work with during the threatcasting process. A pitfall to remain aware of is the analyst's lack of collaboration and the ability to bounce ideas off of other practitioners. Threatcasting by yourself has pitfalls; therefore, it may be useful to share findings to get feedback.

When might an individual Threatcasting be appropriate to use? The analyst typically performs an individual threatcasting session as a way to explore a specific problem set. During the solo exploration, the data may later be used for a small or large group Threatcasting Workshop. Individual Threatcasting is also a useful follow-up to a small or large group Threatcasting Workshop to explore further an outlier topic/threat/concept brought up in the larger workshop.

When conducting an individual Threatcasting session, it is important to understand and be able to articulate the expected output format to the eventual audience. It is easy for scope creep to occur since the analyst is doing this alone. Therefore, stay mindful of this. Additionally, the analyst can get distracted during the process. Staying focused on the Threatcasting Foundation is important in this application. The analyst can always go back later and adjust the research question and rerun the threatcasting as it can yield different results or results that might challenge the previous outcome.

11.2   PHASE 0

Preparation and Curation

For a refresher on Phase 0 of the Threatcasting Method, visit Chapter 2.

11.2.1   DEVELOP THE THREATCASTING FOUNDATION

For an individual Threatcasting, the specific definition of the Threatcasting Foundation is important. Typically, an individual Threatcasting is designed to solve a very specific, singular problem. The analyst will be challenged to constrain the research question so that it is accomplishable. Often an individual Threatcasting is used following a small or large group Threatcasting Workshop. The analyst starts with the findings from the small or large group and uses the individual Threatcasting to explore a single threat or topic in more detail.

Action

Write down your Threatcasting Foundation.

- Can it be constrained further?

- What is the expectation from the Application Areas?

11.2.2   ASSEMBLE THE TEAM

Just because an analyst is performing an individual Threatcasting workshop doesn't automatically mean that there will not be a core team, steering committee, or other analyst(s) involved in the process.

Core Team

When conducting an individual Threatcasting Workshop, the analyst should consider pulling together a core team to gather different perspectives and views on the Threatcasting Foundation, workshop materials, and results. If this individual Threatcasting is an extension of a group Threatcasting, then potentially you can use the same core team or that core team might be able to give feedback on who should be on this new core team.

Action

Answer the following questions.

- Who could give you different perspectives on your topic?

- Can you find a core team that will help to curate the prompts?

Steering Committee

Just like the core team, the steering committee can be a valuable tool for the analyst to connect with SMEs and also for them to review/amplify the results. If this is an extension of a previous group Threatcasting, then you might be able to re-use the previous steering committee.

Action

Brainstorm and answer the following questions.

- Do you need a steering committee? What would they provide that you don't already have access to for this individual Threatcasting?

- Who might be a steering committee member that could help with the identification of prompts and SMEs?

- Is there a person or group that would be strategically helpful to review and then amplify the results?

Additional Analyst(s)

For the individual Threatcasting, the analyst is also the lone participant. It can be helpful to have a "second set of eyes" to review the EBM and post analysis. Just as with the other Threatcasting Workshops, understanding both the expertise and bias of the analyst can help program for better results.

Action

Answer the following questions.

1. Is there a skill or domain area that you wish you had more experience in (as the lone participant/analyst)?

2. Is there another analyst that might help augment that?

3. During the post analysis, if you identify a blind spot or hole in your findings, what will be your plan to bring in another analyst?

Stories from the Lab

<u>Sometimes Validation Comes Late</u>

There comes a point during every Individual Threatcasting that the analyst feels lost. It's only natural. You're all by yourself, working through the prompts, SME interviews, RSW, and EBMs. The post analysis can seem endless with no one to bounce ideas off of.

This happened on a future of supply chain individual Threatcasting that was conducted in the lab. After a flurry of SME interviews, the analyst worked in isolation for two months. Finally, the findings and technical report were complete.

The analyst brought in a secondary analyst to comment and validate the findings. This review helped the lead analyst feel a little better about the work, but not completely.

It was only until the lead analyst started to socialize the report with the SMEs and core team and started doing wider executive summary briefings and seminars that they gathered feedback and new perspectives. The feedback and questions the lead analyst received from these briefings actually evolved the final report, pushing the need for a later revised edition that better clarified some of the report's findings.

During an individual threatcasting, it can be hard sometimes to see where inputs and feedback might come. It's important to stay open and curious at every phase of the method.

11.2.3 SELECT AND GATHER RESEARCH PROMPTS

The curation and selection of the prompts for an individual Threatcasting Workshop has a unique importance. For a small or large group Threatcasting Workshop, the analyst(s) select prompts to give the participants research and insights to explore the Threatcasting Foundation. However, in an individual Threatcasting the analyst will also be the participant. Therefore, the analyst can use the selection and curation of the prompts to challenge their biases and to force multiple perspectives that could affect the Threatcasting Foundation.

The prompts selection can help the analyst obtain better results from the Threatcasting by programmatically forcing topics areas, conflicting ideas, and alternative views.

The process of selecting and gathering the research prompts is discussed more in depth in Sections 2.1.3, 9.2.3, and 10.2.3.

Action

Once you have selected one or more prompts, try to find a counter intuitive prompt or opposing idea that challenges these prompts.

1. Contact an SME through your network and connections, schedule, and record the topic's conversation.

11.2.4 DRAFT THE THREATCASTING WORKBOOKS

When drafting the RSW, there is no significant difference between executing this activity for a group Threatcasting vs. an individual Threatcasting; the same information and analysis is needed on the prompts. Therefore, see Sections 3.1.2 and 9.2.5 for more details.

When drafting the TCW, there is a unique difference between the individual Threatcasting and the group Threatcasting workbook. For the small and large group Threatcasting Workshops, the analyst is preparing the EBM for other people. Refer to Sections 4.1.2 and 9.2.5 for a refresher on this. For the individual Threatcasting Workshop, the analyst is doing the work for themselves. Therefore, you might add more advanced questions to define the EBM.

Drafting an EBM that has challenging or more well-rounded experience and enabling questions can expand the perspectives and depth of the results for an individual Threatcasting. Because the analyst will be able to delve into more details, some more specific questions can be helpful.

Example Individual Threatcasting Experience Questions

- What is the flow of your person's day? What is it like when they wake up? When and where do they experience the event? Are there early indicators of the event? How does their day end?

- What is the first indicator of the event that your person may have missed days before the event happened? What signs did they miss? What went on behind the scenes?

- How and where will the event affect the person's family and larger network across the state/nation/globe? How will it be different?

Example Individual Threatcasting Enabling Questions

- How long before the event did investment in the threat begin? Was the funder aware of the investment?

- What two technologies came to market that unintentionally enabled the threat?

- What policies or regulations were NOT put into effect that would have prevented the threat?

Action

Think about these questions as you draft the TCW.

1. Is there a question that can be added to the experience or enabling questions that would challenge your assumptions?

2. Is there a question that could take your results in a different direction?

11.3   PHASE 1

To open Phase 1, there is no elaborate opening session or welcome. Instead grab a cup/glass of your favorite beverage and get started.

11.3.1 LOGISTICS

With a group of only one, the actual setting to conduct the Threatcasting session becomes vital. The analyst must set up in a private location where they will be uninterrupted for hours. Use a timer to move quickly and not linger longer than is necessary on a single area. The analyst should have all the things needed to seal off from the rest of the world for several hours. This will probably include the following:

- a closed-door room;

- a note on the closed-door "Do Not Disturb please, I am Threatcasting;"

- a computer;

- office supplies such as whiteboards, sticky notes, stacks of paper, and pens/markers;

- caffeine, water, snacks;

- link to SME prompts;

- third-party research to review;

- problem set notes;

- a clock;

- a timer;

- patience;

- ....

Timing

With an individual Threatcasting, it is essential to time the session. Having a clock and timer near the analyst workstation will prove valuable as they help keep the results focused, and it also gets you quicker to the raw data so that you can do post-analysis. Setting a timer for each activity in Phases 1, 2, and 3 will be very helpful.

Bias

In the research synthesis phase, there is a hazard to note when it comes to the individual workshop. Because the analyst has no way of keeping bias in check, it can help to state biases clearly at the beginning. Then state it again at the end of the research synthesis phase to investigate who else should be a part of the discussion, such as subject matter experts, who could help counter bias.

Breaks

Once the analyst has completed a phase in the Threatcasting session—take a break! Stepping away is important. This time gives the analyst not just a mental and physical break, but also a time to reflect. Additionally, the post analysis should take place a day or two after Phase 1, 2, and 3 are conducted.

 NOTE: use previously referenced literature on clearing out your brain and cognition and why taking breaks and shifting in your brain is important.

Action

1. Just like for the group Threatcasting, create a schedule for the individual Threatcasting activities. Time bounding or constraining will help you keep focus.

2. Prepare the space that you will use to conduct the individual Threatcasting.

11.3.2 PRESENTATION OF THE PROMPTS

For a refresher on Phase 1 of the Threatcasting Method, visit Chapter 3.

 There is no difference in the presentation of the prompts between individual Threatcasting and group Threatcasting. See Sections 3.1.1 and 9.3.2 for more details and options.

11.3.3 DATA CAPTURE

Even though the analyst is also the participant, it is important to take the time to review the prompts individually and fill out the RSW. Following the spirit of the RSW gives the analyst time to reflect and capture their thoughts and options on the prompts. As with the small and large group Threatcasting Workshops, it also allows the analyst to identify possible bias and record it in the RSW.

11.3.4   PREPARE FOR NEXT PHASE

Unlike with the group Threatcasting, there is no need to consolidate all the data points on various RSWs into a consolidated RSW to use for Phase 2. That is because only one RSW is created.

However, this is a good point to get up for a walk or some other relaxing activity to allow your brain to reset before starting Phase 2. Depending on your timeline for the individual Threatcasting, you might want to wait a day before proceeding to allow the thoughts to settle.

11.4   PHASE 2 AND 3

For a refresher on Phases 2 and 3 of the Threatcasting Method, visit Chapters 4 and 5.

11.4.1   CAPTURE THE RAW DATA

Guided by the prompts and research synthesis, the analyst engages in a design session to envision a threat ten years in the future. The analyst moves from the high-level macro view of the research to the micro perspective of a person experiencing a threat. To do this:

- use a pseudo-random method to select the data points;

- follow the SFP Process; and

- follow the Experience Design Process.

These processes will generate a qualitative EBM.

Once the experience and enabling questions are answered about "a person, in a palace, experiencing a threat," the analyst begins backcasting by developing a time-phased, alternative-action definition (TAD) phase that generates specific actions that can be taken to disrupt, mitigate, and recover from the threat. Additionally, the analyst identifies the indicators (flags) over the next decade to show that the threat is beginning to manifest and become a reality.

In general, there is nothing different for how to conduct Phases 2 and 3 for an individual Threatcasting or a group Threatcasting event. Of course, logistically it is easier given the reduced number of participants and the fact that the lone participant is also the author of the TCW—so you should not expect them to type in the wrong field or on the wrong tab (as you will see initially in the group Threatcasting events). However, just like in the group Threatcasting events, you should periodically remind yourself to add "more details" to the workbook and not just leave it in your head.

11.4.2   RINSE AND REPEAT

When you are running an individual Threatcasting Workshop it is important to repeat Phases 2 and 3 multiple times. Because the analyst is also the participant, you will need to go through these phases multiple times (with new data points) to create enough raw data to perform the post analysis.

This need to create multiple EBMs on your own allows you to change your threat futures, focusing on different aspects of the prompts to answer the Threatcasting Foundation. You can focus on different threat actors, different locations, and different exploration of the research in the prompts. We often try to develop opposing threat futures. These are threat futures that draw from prompts that might be contradictory. Reconciling these opposing futures can make the post analysis and findings much richer.

Action

Review all the EBMs created.

1. Try to identify what you have missed, especially in the gates and milestones. As the lone participant you might not have all the information you need to know what steps can be taken to disrupt, mitigate, and recover from the threat.

2. Did you cover all the potential aspects of the Threatcasting foundation, or do you need to create one more EBM?

3. Make a note as to places where you might want to do further research, follow up with the SMEs to get their perspective, as well as engage with your core team, steering committee, or additional analysts. This is also typically the place where the lone analyst discovers that they might need to pull in an additional analyst for the post analysis.

4. Start a list of people who you would like to review the findings.

11.5   PHASE 4

For a refresher on Phase 4 of the Threatcasting Method, visit Chapter 6.

After the workshop's conclusion, the analyst studies the raw data, using multiple techniques to cluster and identify possible threats.

A Constrained and Fixed Data Set

It is helpful during post-analysis to only analyze the words that are on the page. As the analyst does the synthesis and clustering, only use the data captured in the workbooks. One pitfall to doing an individual Threatcasting session is that the analyst will often make changes to language midstream because they know the intent in the meaning behind the words instead of just focusing on the words captured.

The raw data sets generated by the Threatcasting Method are constrained and fixed once the workshop is complete. This helps the analyst keep their research within the scope of the Threat-casting foundation, not making any changes after the workshop. But it is also important because the raw data sets and findings can be used in the future by other analysts. The workbooks and post analysis documentation should be 100% transparent so that future analysts and students can use the raw data sets, findings and technical report for other research.

11.5.1   ANALYST(S)

As has been stated previously, the lone participant is also the lead analyst in an individual Threatcasting.

However, it is still very helpful to create a timeline with milestones for this post analysis phase. This will help keep the analyst on track. It is also important to determine your musical playlist for this phase of Threatcasting. Potentially, you will want to use a different playlist than what you used to conduct Phases 2 and 3.

Finally, the analyst will need to prepare their post-analysis workbook in the same manner as if this was a group Threatcasting. See Section 6.1.3 for details and the Project for examples.

11.5.2   PREPARE THE RAW DATA

As the lone participant is also the lead analyst, they will typically (subconsciously) not provide any PII that will need to be removed nor any duplicated or poor-quality data in the fields. Therefore, there SHOULD not be a lot of activities that need to be conducted under the banner of preparing the raw data for analysis. However, the analyst should always double-check before moving on.

Additionally, make sure to create a back-up copy of the RSW and TCW. This should be saved in an appropriate location.

11.5.3   PERFORM THE ANALYSIS

The analyst reviews the multiple EBMs and begins to move through the three phases of post analysis: Summary, Meaning, and Novelty. As the analyst moves through each phase they continually refer back to the Threatcasting Foundation to answer the research question and meet the needs of the Application Areas. See Sections 6.1.3 and 9.5.3 for more details.

Action

1.   Review your notes to see if you need to engage another analyst to review the raw data.

2.   Do you need to reconnect with an SME or do further research to explore the threat futures and meet the needs of the application areas?

11.5.4   GENERATE, VALIDATE, AND REVIEW THE FINDINGS

After the analyst has done the first pass on the analysis and looked at the summary, the next step is to question the results. In this phase, it's essential to write down the research objective so that the analyst is continually seeking and holding oneself accountable to that goal. The analyst can also use subject matter experts as a sort of internal peer review.

Findings

Your findings should be well-documented and can be peer-reviewed by your core team, steering committee, and/or SMEs. The findings are meant to answer the research question and inform the application areas. Additional research can be conducted (if needed) and the final technical documentation should capture the threats, actions, and indicators.

Writing the findings and focusing on implementing those findings tied explicitly to understanding the analyst research question is vital. These findings might also require that the analyst go back to the subject matter experts and interview them again, bouncing the findings off of them.

11.6 PHASE 5

For a refresher on Phase 5 of the Threatcasting Method, visit Chapter 7.

The final phase of the individual Threatcasting workshop translates the findings into an output. This output is pre-determined by the Threatcasting Foundation in Phase 0. The correct output could be a technical report, an academic paper, a podcast, or a sci-fi prototype. The outcome is determined by the person or organization that will be applying or using the Threatcasting Findings.

There is no difference between the types of output for an individual Threatcasting versus a group Threatcasting event. Therefore, see Sections 9.6 and 10.6 for more details on types of outputs.

Stories from the Lab

Always be BLUF-ing

In the lab one practice we have been using for years is to "Start with the BLUF."[20] When the analyst is working on their findings and they want to share it with others in the lab we generally start by asking, "What's your BLUF."

Now this is not the final BLUF. This is just the BLUF as it stands at that specific moment in time. For analysts, BLUFs evolve as their understanding and opinion about the results evolves as well. In this way the analyst is always BLUF-ing, thinking about how they would explain the current state of the project in a paragraph or less.

Business schools use a similar approach. They call it "the elevator pitch". An elevator pitch is the explanation of your new business idea that you are working on if you had just 90 seconds to pitch it to someone in an elevator.

---

[20]   BLUF = Bottom Line Up Front.

It's an effective tool because it forces the analyst to strip down their ideas to be as succinct and understandable as possible.

Even though an analyst might always be BLUF-ing they will generally know when their BLUF is complete. A BLUF is good when the analyst feels that it captures the findings sufficiently and also gives the analyst a starting place when the audience might want to know a little more about the research.

So we tell the analysts to always be BLUF-ing...it's a great place to start.

Action

Practice writing a BLUF and storytelling.

1. Write a half-page, bottom-line up-front summary, highlighting the high level details.

2. Record an audio file of your BLUF. By recording yourself and listening, you can hear if the story is compelling. If it feels flat, revisit writing it and try it again.

This practice of writing and listening to it spoken will help clarify what you are hoping to express. Rewrite and re-record until you feel the BLUF is solid.

CHAPTER 12

Conclusions

Looking Back to Look Forward

Brian David Johnson

"To be honest I didn't think I was going to make it," Nikhil smiled sheepishly.

"What do you mean?" I asked.

"Well Threatcasting can be pretty intense," he continued. "When you are in the middle of it you have a lot going on. There's the core team and the steering committee and the other analysts—you kind of feel the pressure. Especially just before the workshop. You want everything to go right...and because it was my first time...I wasn't sure if I'd be able to do it."

"But you did do it," I prodded Nikhil. His Threatcasting project exploring the future of biological big data had been a big success.

"But I didn't know that," he answered.

"I told you that you could do it," I reminded him. "If you just stick to the method you'll be fine."

"I know you told me that," Nikhil chuckled. "You told me that a lot. I guess I didn't believe you." He laughed again. "But I do now. Threatcasting was great because it let me tie together all of my research and all of the things that I'm passionate about. And I got to meet so many people. That was probably the best past.... meeting all the people and getting to really dig into interesting conversations."

***

As we wrap up the textbook, we thought it would be interesting to go back and talk to students and practitioners who have used Threatcasting in the past. They were all new to it once. We wanted to know what they thought about it. We also wanted to find out what advice they would give you as you start Threatcasting.

***

Captain Eric J Bonick from the United States Air Force is one of the early practitioners of the Threatcasting Method. Bonick is a Deputy Branch Chief at the Air Force Nuclear Weapon Center. But nearly a decade ago he was Cadet Bonick at the United States Air Force Academy (USAFA) in Colorado Springs, Colorado.

USAFA was the first place BDJ taught the Threatcasting Method as he transitioned it from strictly private industry high-tech modeling to broader usage. Bonick and BDJ worked together for years Threatcasting everything from a large group exploring the future of wildfires with the Federal Emergency Management Agency to early cyber threats to national security.

One aspect of Bonick's work that was really interesting was that he ran his own small group Threatcasting with a group of his peers at USAFA looking at a range of possible and probable threat futures in conjunction with nuclear proliferation. In 2021 Bonick and BDJ are back at USAFA, training up a whole new generation of cadets to use Threatcasting for leadership and national security.

He asked Bonick what advice he might have for analysts as they begin using the Threatcasting Method.

"Have an open mind because you will never have all the answers," he replied quickly. "Also cage people's expectations within the realm of reality and the parameters you have set and then let people's creativity take over."

"What do you think is the biggest misconception about the method?" BDJ asked. "Where do people go wrong early?"

"Threatcasting is not predicting the future," he explained. "It is a methodology to help you achieve the future you want and avoid the one you don't want. The future is created by human interactions. The better you can understand those interactions and the ripple effects of them, the more detailed the roadmaps of the future will be, thus making it easier for you to navigate.

"This method is not something you perform once and stick on the shelf. These futures must be routinely evaluated to ensure the path you are on is the one you desire. Course corrections are a natural progression of any good plan."

"You know you've done a large group, small group, and individual Threatcasting," BDJ said. "As a person who has used Threatcasting to address multiple subjects and with a wide array of participants. What surprised you?"

"During the post analysis, you will be amazed by the common themes you will find among so many independent thoughts," he replied. "With the future being dependent on human interactions, if enough people have the same ideas of what tomorrow brings, the odds are good that those ideas will become reality.

"Also, what I have noticed in regard to how most long-term strategic planning is conducted, we come up with the worst case scenarios and we plan against it. Many times, it's ineffective because what ends up happening through those actions yields the very thing we wanted to avoid. What Threatcasting does is not only determine the path you want to avoid but also identify the path you want to achieve. If you set a goal and work toward it while stemming off the things you don't want, you'll be much more successful. Most predictions end up being self-fulfilling prophecies, so why not fulfill the prophecy you want to achieve?"

Poem

Dreams

Hold fast to dreams
For if dreams die [21]

Langston Hughes

Why This Poem Matters

Dreams are an important part of Threatcasting. We look into and explore these dark futures so that we can make the future safer and better for people. Indeed, many of these features can be seen as nightmares, but we must always remember as Hughes writes to "Hold fast to dreams."

Actively working together with people to make the future better is what will get us through the darkness and allow us to dream a different dream for our future.

---

21   Read the full text at https://bit.ly/3m8pGL0.

List of Acronyms

AAR	After Action Report
ACI	Army Cyber Institute
BLUF	Bottom-Line Up-Front
CV	Curriculum Vitae
EBM	Effects-Based Model
EBO	Effects-Based Operations
HCI	Human Computer Interface
IT	Information Technology
PII	Personally Identifiable Information
RSW	Research Synthesis Workbook
SFP	Science Fiction Prototype
SME	Subject Matter Expert
TAD	Time-Phased, Alternative-action Definition
TCW	Threatcasting Workbook
USMA	United States Military Academy
UX	User Experience
WMD	Weapons of Mass Destruction

References

AppliedLogix. (n.d.). *What is a Phase 0 Project for embedded systems development?* Retrieved from https://appliedlogix.com/article/what-is-a-phase-0-project. 13

Army Cyber Institute. (2018). *Dark Hammer: A Retrospective of Science Fiction Prototyping*. Retrieved from https://threatcasting.asu.edu/Dark_Hammer_Retrospective. 208

Atkinson, R. C. and Shiffrin, R. M. (1968). Human memory: A proposed system and its control processes. In *Psychology of Learning and Motivation* (Vol. 2, pp. 89–195). Academic Press. DOI: 10.1016/S0079-7421(08)60422-3. 44

Basso, D. and McCoy, N. (1996). *Study Tools: A Comprehensive Curriculum Guide for Teaching Study Skills to Students with Special Needs.* Twins Publication, Columbia, SC. https://eric.ed.gov /?id=ED412700. 96

Beech, B. (1997). Studying the future: a Delphi survey of how multi-disciplinary clinical staff view the likely development of two community mental health centres over the course of the next two years. *Journal of Advanced Nursing*, 25(2), 331–338. DOI: 10.1046/j.1365-2648.1997.1997025331.x. 40

Bennett, M. and Johnson, B. D. (2016). Dark future precedents: science fiction, futurism and law. In *Intelligent Environments* (Workshops) (pp. 506–513). 52

Birko, S., Dove, E. S., and Özdemir, V. (2015). A delphi technology foresight study: Mapping social construction of scientific evidence on metagenomics tests for water safety. *PloS One*, 10(6), e0129706. DOI: 10.1371/journal.pone.0129706. 40

Bradberry, R. (1950). August 2026: There will come soft rains. In *Martian Chronicles*. New York, NY: Simon and Schuster Paperbacks. 143

Bradfield, R. M. (2008). Cognitive barriers in the scenario development process. *Advances in Developing Human Resources*, 10(2), 198–215. DOI: 10.1177/1523422307313320. 29

Brown, A. (2017, April 18). Why should practitioners publish their research in journals? Retrieved from https://researchforevidence.fhi360.org/practitioners-publish-research-journals. 107

Brown, J. (2021). Thinking Like a Futurist: Investigating the Theories and Processes of Threatcasting Post-Analysis [Doctoral dissertation, Arizona State University]. 130, 132, 134, 135, 137, 138, 139, 142

Cabrera, D., Colosi, L., and Lobdell, C. (2008). Systems thinking. *Evaluation and Program Planning*, 31(3), 299–310. DOI: 10.1016/j.evalprogplan.2007.12.001. 67

Cole, A. and Singer, P. W. (2015). *Ghost Fleet: A Novel of the Next World War*. Eamon Dolan/Houghton Mifflin Harcourt. 56

Cole, A. and Singer, P. W. (2020). *Burn-In: A Novel of the Real Robotic Revolution*. Mariner Boooks. 56

Cowan, N., Elliott, E. M., Saults, J. S., Morey, C. C., Mattox, S., Hismjatullina, A., and Conway, A. R. (2005). On the capacity of attention: Its estimation and its role in working memory and cognitive aptitudes. *Cognitive Psychology*, 51(1), 42–100. DOI: 10.1016/j.cogpsych.2004.12.001. 44

Defense Acquisition University. (n.d.). *DOTmLPF-P Analysis*. Retrieved from https://www.dau.edu/acquipedia/pages/ArticleContent.aspx?itemid=457. 76

Delbecq, A. L., Van de Ven, A. H., and Gustafson, D. H. (1975). *Group Techniques for Program Planning: A Guide to Nominal Group and Delphi Processes*. Scott, Foresman. 40

Desjardins, J. (2017, September 25). Every Single Cognitive Bias in One Infographic. Retrieved from https://www.visualcapitalist.com/every-single-cognitive-bias/. 29

Dweck, C. S. and Leggett, E. L. (1988). A social-cognitive approach to motivation and personality. *Psychological Review*, 95(2), 256. DOI: 10.1037/0033-295X.95.2.256. 44

Eccles, J. (1983). Expectancies, values, and academic behaviors. In J. T. Spence (Ed.). *Achievement and Achievement Motives: Psychological and Social Approaches*. New York, NY: W. H. Freeman. 44

Ellis, S. (2008, April 10). *Frameworks, Methodologies and Processes*. Retrieved from http://vsellis.com/frameworks-methodologies-and-processes/. 5, 9, 10

Feynman, R. (2006). *The Feynman Lectures on Physics* [UNABRIDGED]. Audio CD. Basic Books. 66

Gabriel, D. (2011, May 13). *Methods and Methodology*. Retrieved from https://deborahgabriel.com/2011/05/13/methods-and-methodology/#:~:text=Method%20is%20simply%20a%20research,using%20a%20particular%20research%20method. 5

George Mason University Writing Center. (2008, August 8). *How to Write a Research Question*. Retrieved from https://writingcenter.gmu.edu/guides/how-to-write-a-research-question. 15

Glenn, J. (1972). Futurizing teaching vs. futures courses. *Social Science Record*, 9(3), 26–29. 50

Green, K. and Vergragt, P. (2002). Towards sustainable households: a methodology for developing sustainable technological and social innovations. *Futures*, 34(5), 381–400. DOI: 10.1016/S0016-3287(01)00066-0. 76

Höjer, M., Gullberg, A., and Pettersson, R. (2011). Backcasting images of the future city—Time and space for sustainable development in Stockholm. *Technological Forecasting and Social Change*, 78(5), 819–834. DOI: 10.1016/j.techfore.2011.01.009. 76

Inayatullah, S. (2008). Six pillars: futures thinking for transforming. *Foresight*, 10(1), 4-21. DOI: 10.1108/14636680810855991. 53

Incrementalism. (2021, June 10). In *Britannica*. https://www.britannica.com/topic/incrementalism. 85

Joint Chiefs of Staff (2021). *Department of Defense Dictionary of Military and Associated Terms*. United States Government. Retrieved from https://www.jcs.mil/Portals/36/Documents/Doctrine/pubs/dictionary.pdf. 64

Johnson, B. D. (2011). *Science Fiction Prototyping: Designing the Future with Science Fiction*. Synthesis Lectures on Computer Science. Morgan & Claypool Publishers. DOI: 10.2200/S00336ED1V01Y201102CSL003. 52

Johnson, B. D. and Vanatta, N. (2017). What the heck is threatcasting? *Future Tense*. Retrieved from https://threatcasting.asu.edu/sites/default/files/2019-11/what-the-heck-is-threat-casting.pdf. 87

Johnson, B. D., Vanatta, N., Draudt, A., and West, J. R. (2017). The New Dogs of War: The Future of Weaponized Artificial Intelligence. Retrieved from https://apps.dtic.mil/sti/citations/AD1040008. 201

Johnson, B. D. (2018). *Hero*. Army Cyber Institute, West Point. Retrieved from: https://threatcasting.asu.edu/graphic-novella/hero-2028. 117

Karray, F., Alemzadeh, M., Abou Saleh, J., and Arab, M. N. (2008). Human-computer interaction: Overview on state of the art. *International Journal on Smart Sensing and Intelligent Systems*, 1(1), 137–159. DOI: 10.21307/ijssis-2017-283. 50

Kolb, D. A. (2014). *Experiential Learning: Experience as the Source of Learning and Development* (ed. 2). FT press. 44

Kyle, C. (2008). *RMA to ONA: The Saga of an Effects-based Operation* [Monograph, School of Advanced Military Studies, United States Army Command and General Staff College]. DTIC. https://apps.dtic.mil/sti/pdfs/ADA499725.pdf. DOI: 10.21236/ADA499725. 63

Landeta, J. (2006). Current validity of the Delphi method in social sciences. *Technological Forecasting and Social Change*, 73(5), 467–482. DOI: 10.1016/j.techfore.2005.09.002. 40

Lindblom, C.E. (1959). The science of "muddling" through. *Public Administration Review*, 19(2): 79-88. Retrieved from https://www.jstor.org/stable/973677. DOI: 10.2307/973677. 85

Linstone, H. A. and Turoff, M. (2011). Delphi: A brief look backward and forward. *Technological Forecasting and Social Change*, 78(9), 1712–1719. DOI: 10.1016/j.techfore.2010.09.011. 40

Lovin, A. B. (1976). Energy strategy: the road not taken. *Foreign Affairs*, 55, 65–96. DOI: 10.2307/20039628.

Markowitz, E. (2020, February 6). *How to Write an Executive Summary*. Retrieved from https://www.inc.com/guides/2010/09/how-to-write-an-executive-summary.html. 108

Maslow, A. H. (1943). A theory of human motivation. *Psychological Review*, 50(4), 370. DOI: 10.1037/h0054346. 44

Meissner, P. and Wulf, T. (2013). Cognitive benefits of scenario planning: Its impact on biases and decision quality. *Technological Forecasting and Social Change*, 80(4), 801–814. DOI: 10.1016/j.techfore.2012.09.011. 29

MindTools. (n.d.). *Paraphrasing and Summarizing*. Retrieved from https://www.mindtools.com/pages/article/paraphrasing-summarizing.htm. 96

Nenniger, P. (1992). Task motivation: An interaction between the cognitive and content-oriented dimensions of learning. In K.A. Renninger, S. Hidi, and A. Krapp (Ed(s).). *The Role of Interest in Learning and Development* (pp. 121–150). New York, NY: Psychology Press. 44

Ontology (information science). (2021, April 10). In Wikipedia. https://en.wikipedia.org/wiki/Ontology_(information_science). 5

Perception Institute. (n.d.). *Implicit Bias*. Retrieved from https://perception.org/research/implicit-bias/. 29

Project Implicit. (n.d.). *Project Implicit*. Retrieved from https://www.projectimplicit.net/. 29

Renninger, K. A., Hidi, S., Krapp, A., and Renninger, A. (Eds.). (2014). *The Role of Interest in Learning and Development*. Psychology Press. DOI: 10.4324/9781315807430. 44

Rickerman, L. (2003). *Effects Based Operations: A New way of Thinking and Fighting* [Monograph, School of Advanced Military Studies, United States Army Command and General Staff College]. DOI: 10.21236/ADA416050. 52

Ringland, G. and Schwartz, P. P. (1998). *Scenario Planning: Managing for the Future*. John Wiley and Sons. 55

Robertson, T. and Simonsen, J. (2012). Challenges and opportunities in contemporary participatory design. *Design Issues*, 28(3), 3–9. DOI: 10.1162/DESI_a_00157. 50

Robinson, J. B. (1982). Energy backcasting: A proposed method of policy analysis. *Energy Policy*, 10(4), 337–344. DOI: 10.1016/0301-4215(82)90048-9. 75

Robinson, J. B. (1990). Futures under glass: a recipe for people who hate to predict. *Futures*, 22(8), 820–842. 76

Rowe, G. and Wright, G. (1999). The Delphi technique as a forecasting tool: issues and analysis. *International Journal of Forecasting*, 15(4), 353–375. DOI: 10.1016/S0169-2070(99)00018-7. 40

Shedroff, N. (2001). *Experience Design 1*. New Riders Publishing. 51, 70, 73

Shonka, M. and Kosch, D. (2002). *Beyond Selling Value: A Proven Process to Avoid the Vendor Trap*. Dearborn Trade Publishing. https://www.amazon.com/Beyond-Selling-Value-Proven-Process/dp/0793154707. 109

Stalter, A. M., Phillips, J. M., Ruggiero, J. S., Scardaville, D. L., Merriam, D., Dolansky, M. A., Goldschmidt, K. A., Wiggs, C. M., and Winegardner, S. (2017). A concept analysis of systems thinking. *Nursing Forum*, 52(4), 323–330. DOI: 10.1111/nuf.12196. 68

Staples, W. G. (Ed.). (2007). *Encyclopedia of Privacy: AM* (Vol. 1). Greenwood Publishing Group. 93

Tuominen, A., Tapio, P., Varho, V., Järvi, T., and Banister, D. (2014). Pluralistic backcasting: Integrating multiple visions with policy packages for transport climate policy. *Futures*, 60, 41–58. DOI: 10.1016/j.futures.2014.04.014. 76

Tversky, A. and Kahneman, D. (1982). Judgement under uncertainty: heuristics and biases. In Kahneman, D., Slovic, P. and Tversky, A. (Ed(s).). *Judgement Under Uncertainty: Heuristics and Biases* (pp. 3–10). New York, NY: Cambridge University Press. DOI: 10.1017/CBO9780511809477.002. 29

Van der Heijden, K. (2011). *Scenarios: The Art of Strategic Conversation* (ed. 2). John Wiley and Sons. 50

van de Kerkhof, M., Cuppen, E., and Hisschemöller, M. (2009). The repertory grid to unfold conflicting positions: the case of a stakeholder dialogue on prospects for hydrogen. *Technological Forecasting and Social Change*, 76(3), 422–432. DOI: 10.1016/j.techfore.2008.07.004. 76

Weller, S. C., Vickers, B., Bernard, H. R., Blackburn, A. M., Borgatti, S., Gravlee, C. C., and Johnson, J. C. (2018). Open-ended interview questions and saturation. *PloS One*, 13(6), e0198606. DOI: 10.1371/journal.pone.0198606. 41

Wicks, D. (2010). Coding: axial coding. In Mills, A. J., Durepos, G., and Wiebe, E. (Ed(s).). *Encyclopedia of Case Study Research* (Volume 1, pp. 153–155). Thousand Oaks, CA: SAGE Publications. 95

Authors' Biographies

Brian David Johnson is a Futurist, Author, and Professor of Practice at Arizona State University's School for the Future of Innovation in Society. In his private practice he works with governments, militaries, trade organizations, and startups to help them envision their future. He has over 40 patents and is the author of a number of books of fiction and nonfiction, including *The Future You, WaR: Wizards and Robots, Science Fiction Prototyping, Humanity and the Machine: What Comes After Greed?*, and *Vintage Tomorrows*. His writing has appeared in publications ranging from *The Wall Street Journal* and *Slate* to *IEEE Computer* and *Successful Farming*, and he appears regularly on The BBC, Bloomberg TV, PBS, Fox News, and the Discovery Channel. He has directed two feature films, and is an illustrator and commissioned painter.

Natalie Vanatta is a U.S. Army cyber officer and currently serves as the Emerging Technologies Research Team Lead at the Army Cyber Institute (the Army's think tank on cyber). Here she focuses on bringing private industry, academia, and government agencies together to explore and solve cyber challenges facing the U.S. Army in the next 3–10 years in order to prevent strategic surprise. She holds a Ph.D. in applied mathematics as well as degrees in computer engineering and systems engineering. She is an adjunct Associate Professor at Columbia University as well as an Academy professor at the United States Military Academy at West Point.

Spanning a military career of over 20 years, Natalie has served and operated around the globe (on three continents) in order to defend the nation. She was a platoon leader in Germany, a company commander in Kuwait, and a Battalion S3 (operations officer) in Italy. LTC Vanatta has also served as a Distinguished Visiting Professor at the National Security Agency (in both 2017 and 2021), the technical director of Joint Task Force Ares, and a Team Leader within the Cyber National Mission Force focused on defending critical infrastructure for the nation.

Cyndi Coon is the CEO, and Founder of Laboratory5 Inc., where she builds frameworks and human ecosystems to provide decision-makers in government agencies, military, higher education, NGOs, and nonprofits with scenarios they need to take action. Cyndi is the Chief of Staff at the Threatcasting Lab, Producer at Applied Futures Lab, Producing Director at the Weaponized Narrative Initiative, and a producer for the president's office at Arizona State University. Through these roles, she engages with multiple audiences across numerous mediums to create narratives for tools, training, and applied outcomes and she is a trained facilitator. Topics she has created include: Artificial Intelligence, Big Data, Coolabilities, Dual-Use Technologies, Ecosystems, Education, Experiential Events, Future of Sports, Gender, Information Shaping, Information Disorder, Justice, Moral Injury, STEM, Space, Threatcasting, Weaponized Narratives, and Covid-19. Cyndi is a subject matter expert in Information Shaping, Creativity, and Threatcasting. Trained as an artist, Cyndi received a Masters Degree of Fine Art from Arizona State University, Tempe, Arizona and a Bachelors of Fine Art Degree from Kendall College of Art and Design in Grand Rapids, Michigan. Cyndi taught at Arizona State University and in the Maricopa Community College system.

DISCLAIMER

This publication was privately produced and is not the product of an official of the United States Army acting in an official capacity. The contents of this publication, including words, images, and opinions, are unofficial and not to be considered as the official views of the United States Military Academy, United States Army, or Department of Defense. Neither this publication nor its content are endorsed by the United States Military Academy, United States Army, or Department of Defense.

Printed in the United States
by Baker & Taylor Publisher Services